The Introduction of the Merino Sheep To Europe

by C.P. Lasteyrie

with an introduction by Jackson Chambers

Self Reliance Books

Get more historic titles on animal and stock breeding, gardening and old
fashioned skills by visiting us at:

http://selfreliancebooks.blogspot.com/

Introduction

I am pleased to present yet another practical title on breeding and raising livestock.

The work is in the Public Domain and is re-printed here in accordance with Federal Laws.

As with all reprinted books of this age that are intended to perfectly reproduce the original edition, considerable pains and effort had to be undertaken to correct fading and sometimes outright damage to existing proofs of this title. At times, this task is quite monumental, requiring an almost total "rebuilding" of some pages from digital proofs of multiple copies. Despite this, imperfections still sometimes exist in the final proof and may detract from the visual appearance of the text.

I hope you enjoy reading this book as much as I enjoyed making it available to readers again.

Jackson Chambers

DEDICATION.

TO THE RIGHT HONOURABLE

SIR JOSEPH BANKS, BART.

K. B. P. R. S. &c. &c. &c.

SIR,

WHEN submitting to the perusal of British breeders a work, which describes the progress of Merino-sheep in various States, it is natural that I look towards you as its patron; since your exertions, in furtherance of his Majesty's gracious and paternal views, first brought this valuable race to Britain, where it promises to become of such incalculable benefit.

The following pages will amply shew

the success which has attended the naturalization of this breed in countries materially differing from each other; and they will also explain the obstacles, which have, at first, been opposed to a general adoption of this race.

In no country have those difficulties been greater than in our own, as you, Sir, well know; for even your utmost efforts, aided by those of other spirited and enlightened individuals, were unable, during a considerable lapse of years, to dispel the mist of ignorance and prejudice.

Far, however, from allowing this to damp your ardour, you felt confident that the growers of fine wool would ultimately see their own interest in its proper light; and therefore persevered in the course pointed out by his Majesty's wisdom, and supported by his inexhaustible patience for the general benefit of his dominions, till you have spread through the Island that conviction which was impressed upon your own mind above twenty years ago.

The silly opposition to this race, founded on the idea, that Merino *wool* would degenerate for want of the Spanish climate, pasturage, and journeys, has dwindled into insignificance; while the aspersions on the quality of the *mutton*, refuted and crushed by the test of actual experience, have sunk into merited contempt.

The period is, therefore, arrived, when our revered Monarch receives the fruits of his exalted patriotism, and you, Sir, obtain the reward of your labours in the gratifying reflection of having essentially contributed to the foundation of our national independence as to superfine wool; a prospect on which every true Englishman must dwell with unalloyed delight.

DEDICATION.

Wishing you may long enjoy this proud
and enviable sensation, I have the honour
most respectfully to remain,

Sir,

Your very obedient and much
obliged Servant,

BENJ. THOMPSON.

Redhill Lodge, near Nottingham,
Dec. 1, 1809.

TABLE OF CHAPTERS.

PART THE FIRST

PART THE SECOND.

INTRODUCTION

OF

MERINO SHEEP, &c.

PART THE FIRST.

CHAPTER I.

GENERAL REMARKS ON THE NATURALIZATION OF MERINOS.

THE Governments of Europe had long perceived the advantages, which would result to agriculture and commerce from the introduction of fine-woolled sheep into their several states; but their views being opposed by the ignorance and prejudices of the times, many years elapsed before they endeavoured to realize an idea, which had, at first, appeared chimerical.

B

At length, however, men appeared, equally to be admired for their patriotism and knowledge, who have endeavoured with zealous perseverance to enlighten their fellow-countrymen, and prove by facts that nature, far from opposing any exertion to preserve fine-woolled sheep in certain climates, appeared, on the contrary, ready to assist the attempts of industry.

Among those, who have thus deserved well of their country, may be mentioned Alstroemer in Sweden, Koenig in Saxony, Fink in Prussia, Magnis in Silesia, Daubenton and Gilbert in France, Tivent in Holland, and Nelson in Denmark.

Almost a century has elapsed since Sweden introduced and naturalized, in her rigorous climate, the fine-woolled sheep of Spain. During almost fifty years, Norway and other countries, in the North of Europe, have had the same success; yet at the time these nations began their experiment, we attached to them the appellation of *barbarous*. The French people, at that time led away to the pursuits of luxury and pleasure by a corrupt and dissipated court, and suffering themselves to be dazzled by the vain appearances of false grandeur, neglected those studies and labours, which found the prosperity and durability of empires. France, indeed, though particularly favoured by nature, has been, and is still, backward with respect to many objects of industry.

It will be seen by the exposition of facts relative to the improvement of wool, how much time and trouble it has cost us to reach even the situation, in which we at present stand.

The reader, when acquainted with what has passed in foreign countries, will be no doubt surprized that obstacles, so formidable, should have been encountered in trying to prove that, which was known and attested by long experience among different nations, with which we maintained the most intimate political relations. But the people, as I shall several times have occasion to remark in the course of this work, are not sufficiently enlightened, and their rulers have not a sufficient regard for the general good to excite and accelerate improvements, which are really useful.

Several sovereigns, nevertheless, struck with the success which has attended the Merino race in different countries, have, within a certain period, strenuously encouraged this important branch of rural economy, partly by introducing thee fine-woolled sheep into their states, and partly by increasing the number of individuals already naturalized there. In Germany it is observable that this tendency towards perfection becomes more general among the farmers every day.

I conceive that, in a Treatise, which I formerly published, I proved the fine wool of Spain to be neither dependent on the journeys of the sheep, nor the soil, nor the climate, nor the pasturage; but, that this perfection must be ascribed to other causes, and that it is possible to grow in France, and elsewhere, wool of the same quality as that of Spain. My tour in the North of Europe has supplied me with facts and observations, which have confirmed this truth. In the greater number of flocks, which I have there examined, I have found wool, which, judging either by the eye or the touch, equals in beauty and in fineness the piles of Segovia and Leon; whence there can, in my opinion, be no doubt of easily obtaining superfine fleeces in every country of Europe, where the sheep can be supplied with pasturage, and winter-food, adapted to their support. These wools produce cloths as fine, silky, and supple as those manufactured from Spanish wool; a result, proved by the experiments made in France and other countries. But were it true that the nature of the food, climate, and other local circumstances had a certain influence on the intrinsic qualities of wool, such as elasticity, strength, softness, &c., it would be a fact no less established, that cloths can be obtained, at all events, fine and beautiful enough to satisfy those, who are most difficult in these points; and that a nation, therefore, may easily be independent of Spain, while she supplies her fine manufactures with an article grown on her own soil.

Still, as these truths continue to be doubted by some persons, and ought to have a great influence on our agriculture and commerce, I have deemed it a duty to publish such facts as may impart to them a new degree of force, and certainty.

I am about, therefore, to give an account of the Spanish sheep naturalized in Europe, beginning with that country, into which they were first introduced.

CHAPTER II.

SWEDEN.

THERE is, doubtless, just ground for surprise that Sweden, which seems to be a country the most unfavourable in Europe for rearing fine-woolled sheep, should, nevertheless, have been the first to naturalize this valuable race; but it is still more astonishing that men are to be found in France, who deny the possibility of a naturalization, proved thus easy in Sweden for almost a hundred years.

Mr. Alstroemer, who had in 1715 made some experiments with a view to improve the wretched breeds of Sweden, did not think that this improvement should be limited to the good qualities obtained from German and English sheep. Accordingly he imported a flock of Merinos from Spain in 1723; and succeeded in the naturalization and propagation of this race in an austere climate, while it seemed incapable of existing but in a warm country.

The Swedish Government, convinced that ignorance would oppose great obstacles to the preservation and increase of this new breed, established in 1739 a school for shepherds, the direction of which was bestowed on Mr. Alstroemer. In 1740, the States awarded a fund for premiums to those breeders, who sold Spanish rams; and from this date to 1780 a gratuity of 25 per cent. was allowed on the sale of all wool, which was proved to be fine and of good quality[*]. These advantages were reduced in 1780 to 15 per cent.; again in 1786 to 12 per cent.; and in 1792 they were entirely suppressed.

It is ascertained by the exact returns which have been communicated on the part of Mr. Schulzenheim, a member of the Stockholm

---

[*] It may appear to some unnecessary, after calling wool *fine*, that a stipulation should also be made as to its *goodness of quality*; but the best judges of that article in all countries will sanction it. Great Britain affords instances of clothing wool equally fine to all appearance, and yet of qualities so dissimilar as to justify the terms good, bad, and indifferent.

" It may be affirmed," says Mr. Bakewell, " that taking two packs of sorted wool of *the same apparent fineness*, one possessing in an eminent degree the soft quality, the other of the hard kind, the former will, with the same expence to the manufacturer, make a cloth, the value of which shall exceed the latter full 25 per cent."

This subject is fully discussed by the above author in his " Observations on the Influence of Soil and Climate upon Wool," to which I refer the reader.

T.

Academy of Sciences, that the quantity of fine wools grown in Sweden, and sold to the public magazines as well as manufacturers, amounted between the years 1751 and 1790 to 3,402,961 francs, and that government expended during the same period 1,413,450 francs* in the encouragement given to this species of industry. The quantity of fine wool produced in the country was more than this calculation includes; because a part of it was wrought by individuals, who are accustomed, in Sweden, to manufacture their cloths, and other woollen articles in their own houses. Further, the distance of the magazines to which those, who claimed the premiums, were obliged to take the wool, and the successive diminution of those premiums, caused the quantity thither conveyed to be at length no fair criterion of that really grown.

It is to be remarked that the quantity of Spanish wool imported into Sweden during the same space of 40 years was no more than 2,623,040lbs., and that consequently the quantity of fine wool grown in the country, surpassed this importation to the amount of 772,920lbs.

The importation of Spanish wool has sustained

---

* A franc is about eleven-pence sterling.

a new diminution since 1790, though the manufacture of fine cloths has increased.

It will, perhaps, appear surprizing that the Merino sheep, naturalized in Sweden so long, should not yet have increased so much as to supply the kingdom with the quantity of wool, consumed by the superfine manufactures; but exclusive of the political causes, which it would require too much digression from our subject to develope, there are other reasons, both moral and physical, to which this effect may be assigned.

The dealers in wool and manufacturers of cloth have here, as elsewhere, an interest in depreciating the fine wools of the country; and they find it easy to keep them below their real value as articles of commerce. The moderate price too of some German fine wools, and especially those of Eiderstadt (the importation of which is easy) must, of course, operate against the increase of Merinos.

These fine-woolled sheep have been almost exclusively bred by persons in easy circumstances. The peasants of Sweden, who have, in general, only a very small number of sheep, and who are themselves in the habit of manufacturing the articles of attire for their own use, have been compelled to preserve their original breed, as these alone could provide them with the long coarse wool, which is become indispensable to them.

The same prejudices have also arisen in Sweden, which exist among ourselves. Some persons even now think, in defiance of the multiplied facts, which attest the contrary, that the Merino race is not of a nature to prosper in this country, and that it will not make the same advantageous return as the native race. Others, again, fancying that the sheep of this breed may be abandoned to themselves, and that no care whatever is necessary, treat them exactly like the general flocks of the country, or in other words, shut them in close, damp, ill-contrived buildings, the air of which is infected by too great a degree of heat; feed them, during one half of the year, on an insufficient quantity of straw or other bad fodder; and pasture them on the most hungry soils, and bogs, where the animals can find neither the quantity nor quality of support proper for them. It has naturally been observed, that the Spanish race, subjected to this vicious regimen, insensibly degenerated, and that their wool became less fine, less soft to the touch, and deficient in length.

I have seen many flocks of the Merino breed

in Sweden, and I have constantly observed that
those animals always failed with respect to beauty and fineness of fleece, which were under the
care of inattentive people; while in the flocks
of careful breeders I have found sheep, which
preserved their primitive quality of fleece. As
I have made the same remark in Holland, and
various parts of Germany, which I shall hereafter shew; and as the best agriculturists of
these different countries, whom I have consulted, have uniformly attributed the degeneration
of the breed to causes which I shall point out,
let me advise those, who adopt the Merino race,
to guard themselves against the opinion that it
will preserve the excellence of its fleece, if it be
subjected to wrong treatment, or abandoned to
carelessness and ignorance. One or two facts
have been brought forward against this my ar-
gument; but these facts, were they well established, would be unable to overturn the number of others, acknowledged by agriculturists in
various countries. I might, if it were necessary, here quote my Treatise, in which I adduce
many facts to prove that the Merinos degenerate even in Spain, if they have not the necessary care bestowed on them. I have found in
Spain the same results as I witnessed in Sweden.

I have seen, in these two kingdoms, animals of the same breed, and inhabiting the same country, of which some preserved their perfection of fleece, while others produced wool of a very inferior quality. This difference proves that nature, in almost all circumstances, will not lend herself to our wants till we know how to consult her with respect to her operations, follow her in her progress, and second her with our intelligence.

The wish to see a valuable race of animals rapidly propagated may carry the friends of public good too far. It should be remembered that in endeavouring to prove too much, we often prove nothing; and that it is always dangerous to introduce theories, the application of which may militate against the progress of art.

The undistinguished mixture of Merinos, and their crosses with the indigenous breeds, has also contributed in no small degree to the debasement of the fine-woolled breed in Sweden; which might, with care, have been easily preserved in full perfection.

Another obstacle to the increase of the Spanish race exists in the want of food. The support of cattle, and more particularly sheep, is an object of far greater expence in Sweden than in more southern countries; the rigour of the climate making it necessary to keep the animals housed during six or seven months of the year. The farmer ought, therefore, to have abundance of fodder for the support of his herds and flocks during so long a winter; which necessity makes that fodder higher in price, and less easy to be obtained.

Greater profit, however, will result, even under these circumstances, from a Merino flock, than from any other kind: but as fodder of a quality sufficiently good, or in quantity sufficiently large, cannot in some places be procured; and moreover as farmers are loth to abandon their old habits, especially when obstacles are to be overcome; this breed has not been supported in Sweden with all that energy which the wants of the public, and the interest of individuals, rendered necessary.

In 1764, there were 65,369 sheep of the pure Merino race, and 23,384 of the mixed breed, which yielded wool of good quality. It is difficult to determine the present number, because no basis exists, on which a calculation can be formed, since the premiums of encouragement were suppressed; but it is certain, that the number has been continually increasing; and though the manufactures of fine cloths in Sweden are not entirely supplied from the growth of that country, the importation has, nevertheless, been less considerable in latter times, while the superfine fabrics have been increasing. Taking, therefore, the quantity of wool consumed by these manufactures for a criterion, we may suppose that the number of fine-woolled sheep in Sweden, either pure Merino, or possessing some portion of that blood, now amounts to about a hundred thousand. This number, which constitutes a 25th part of all the sheep in the kingdom, must be considered very large, when reference is made to the wants of the nation, and the general state of agriculture in the country.

The Merinos preserve in Sweden their original form. Their fleeces are close and thick. The wool loses nothing in point of fineness, length, or elasticity; and the quantity of it is greater than in Spain, if the animal has had sufficient food. There are instances of rams, whose fleeces have weighed 13 lbs. I have remarked that this naturalized race is larger and stronger than the sheep are in Spain.

On Mr. Schulzenheim's farm, at Gronsoe, in the province of Upland, I found a flock, derived from sheep which had been imported from Spain 55 years before, the wool of which, on a comparison with that of Merinos recently brought from the same country, was not inferior to it either in beauty or fineness.

Mr. Schulzenheim has at six different times imported Spanish rams, for the purpose of imparting, if possible, a further degree of perfection to his flock; and it has been observed that all, except those which arrived in 1778, carried wool inferior to that of the original stock, though they had been skilfully selected under the direction of Mr. Gahn, the Swedish consul at Madrid, who is a nephew of Mr. Schulzenheim.

The same observation has been made in France. The sheep selected in Spain by Gilbert, who was a good judge, have fleeces inferior to those of some at Rambouillet.

I must add, that Mr. Schulzenheim had preserved the sheep imported from Spain long enough to see their descendants to the fifth generation. The comparison of the wool proved that the last descendants had lost none of the qualities, which recommend this race to notice. Such facts shew, in a manner decisive and peremptory, that the Merino breed may be propagated and maintained in cold countries, without losing the fineness or the goodness of fleece.

Among the other fine-woolled flocks, which I saw in Sweden, I must mention, in corroboration of my tenets, that belonging to Baron Macklaw, which had been 20 years on his estate at Edet, 20 leagues north of Gothenburgh.

I was convinced by an examination of these flocks, that their wool possessed all the qualities necessary for the manufacture of fine cloths.

I collected, during my travels, various specimens of wool, which I shall mention in the course of this work. The Agricultural Society in the Department of the Seine, to which I presented them, has decided on the beauty and fineness, by which they are distinguished.

And here, I cannot refrain from pointing out to my readers, the advantages, which would result to agriculture from collections of every kind relating to that pursuit. It is well known that Antiquarian science, that Physic, Chemistry, Botany, Mineralogy, as well as other branches of Natural History, must be allowed to have made their progress through the public and private collections, which have been formed in modern

times. Rural Economy, having for its object
the reproduction of certain beings and certain
substances, as well as the execution of certain
practices and methods; this science being
founded on facts too, as is all human knowledge; it is surely not less important to consoli-
date and submit to view the objects, which confirm these facts; to compare them, to throw
light upon theory, and to give practice a more
prompt as well as more certain direction. If
there had existed, forty years ago, a collection
with a view to Rural Economy such as I now
possess, the improvement in our fine-woolled
sheep would probably have commenced thirty
years before it did; because it would have been
then evident, from an inspection of the specimens which had been grown by Merinos naturalized during several generations in Sweden
and Saxony, that it was very easy to obtain superfine wool in France by introducing the
breeds, which throve in the coldest climates,
and in countries the least favoured by nature.

CHAPTER III.

THE DANISH STATES.

THE sheep of Norway have been improved by the breeds of England and of Spain. A considerable amelioration in the wool of the bailiwick called Smaalchnem is attributed to a Merino ram, which was imported more than fifty years ago on the western coast of Norway; but this improvement has not extended to other parts of the country, and has not produced that degree of perfection, which might have been expected. I know not whether this has arisen from the carelessness of the inhabitants, or from the general bad treatment to which sheep are there subjected. These animals are almost famished during one half of the year. They are kept through the winter in dark and ill-contrived buildings, the heat of which is excessive. The lambs are allowed to copulate at six months old; and the shearing takes place several times in the course of the year, partly at seasons when the loss of covering must be highly detrimental to the animals' health.

The observations, which I was induced to make on the breeds of Norway, and those of

Iceland naturalized in the former country, induce me to offer here some details, which may be interesting to the reader. Individual sheep are to be found in Norway, carrying fleeces, which weigh seven pounds. The flocks, of which a few are black, are in some districts shorn twice, thrice, and even four times in the course of a year. A greater quantity of wool is obtained by this repetition; but it is less adapted to the purposes of manufacture.

These animals endure the most rigorous cold without being injured by it. I have seen, in latitude 64, flocks left in the islands, where they exist in a wild state amidst the snows, without ever receiving food from the hand of man; and they are of so intractable a nature, that they cannot be taken but by pursuit on horseback. They are, in fact, so accustomed to the impressions of the atmosphere, that they cannot endure a sudden transition from a wild state to domestication. Some persons, in the neighbourhood of Gothenburgh, having procured sheep of this kind, housed them during a severe season; but the animals were incapable of existing under the privation of open air, and fell victims to the trial. This fact proves that a free circulation of air is always salutary to sheep, and that the too sudden changes of climate, food, treatment, &c. to which we must under certain circumstances submit, should be properly regulated by art.

The sheep of Faro and Iceland, which appear to have originally come from Norway, live also in a wild state, and endure a climate still more severe. The animals latterly conveyed from these islands to Norway, carry a fleece in which are three sorts of filaments. The first is five to nine inches long; it is stiff, strong, and coarse, of a milky colour and opaque, appearing to be of the same nature as the kemp, or stitchel-hair. The second kind is shorter, viz. four or five inches; it appears to be wool of a medium quality, and reflects a silvery brightness. The third is only two to three inches long, and may, perhaps, be compared, in point of fineness, with the best kind of Spanish wools. It is elastic, pliant, and extremely soft to the touch. I have not had an opportunity of ascertaining the different proportions of these wools, or hairs, to be found in each fleece. I only know that in Iceland a fleece contains five ounces of coarse or kemp wool, twenty-four ounces of ordinary quality, and nineteen ounces of superfine. Those animals of the Iceland breed which I saw, had only a small quantity of the fine and silky sort*.

I found in some districts of Norway, sheep, which carried wool of three sorts similar to those above quoted; and the short silky wool, which they produced, had a right to be ranked among the superfine. It resembles that of the

* " The Icelandic sheep," says Dr. Von Troil (in his Letters, describing that island) " have straight ears standing upright, a small tail, and it is common to meet with sheep, that have four or five horns. In some places they are kept in stables during winter; but they are generally left to seek their food themselves in the fields. It is remarkable that they are fond of hiding themselves in caves, (of which there are a great many in Iceland) in stormy tempestuous weather; but when they cannot find any retreat during a heavy fall of snow, they place themselves all in a heap with their heads to the middle, and bent towards the ground, which not only prevents them from being so easily buried under the snow, but facilitates the owner finding them again. In this situation they can remain several days; and there are examples of their having been forced by hunger to gnaw off each other's wool, which, forming into balls in their stomachs, presently destroys them. They are, however, generally soon sought for and disengaged. There are no wild sheep, as has been pretended by some; for they all have their owners, who keep an exact account of them, and when they are driven to the mountains, they are scarcely ever without a shepherd to attend them."   T.

Shetland sheep; if I may judge by the samples of the latter sent to me from Shetland. The wool of the Shetland sheep is sorted in England, and the best made into stockings, which are generally sold at ten guineas per pair. The fleeces of the Iceland and Norwegian sheep would, if properly sorted, furnish wool as fine and silky as that of Shetland *.

* The only Shetland sheep that I ever saw, were three ewes belonging to that enlightened and patriotic agriculturist, Mr. Tollet of Staffordshire; who was trying the effect of the Merino cross upon them. The wool of these animals was very fine; but the whole fleece was debased by a considerable intermixture of coarse hair. They were shorn in June last, with three Merino-Shetland wethers bred from them. The weight of the three Shetland fleeces together was only *two pounds fourteen ounces;* the weight of the three fleeces from their sons was *thirteen pounds, four ounces.* Exclusive of this great increase in quantity, it is hardly possible to convey an idea of the improvement made in the quality of the wool: I rate it at the lowest when I say that the article was of double value. It appears that there are two varieties of the Shetland breed; the one being furnished with a fleece, which is entirely of a soft and cottony quality; while the other carries a mixture of fine wool and coarse hair. It must be the first of these to which Mr. Lasteyrie refers, when he mentions so high a price for stockings made from the wool; and it must be the last, which I saw at Mr. Tollet's shearing.   F.

The Danes, encouraged by the example of Sweden, procured Spanish sheep from that kingdom about twenty-three years since, and the descendants of this importation exist at the present day, though in very small numbers. Some of these animals have preserved their pristine fineness of wool, while others have degenerated, from causes similar to those already mentioned. I examined a flock of Spanish sheep at Esserum, a royal domain, the animals composing which had been brought from Sweden several years before, and their wool was of good quality. .

In 1797, the Danish government procured three-hundred Merino sheep from Spain, which were placed on the above farm of Esserum, about eight leagues from Copenhagen. This flock is composed of the best Spanish breeds, viz. the Escurial, the Guadeloupe, the Paular, the Infantado, the Montareo, and the Negretti. The *Escurial* breed is accounted, as to fineness of wool, the most perfect of all the travelling flocks in Spain. The *Guadeloupe* sheep are remarkable for symmetry, as well as for the quantity and quality of their wool. The *Paulars* are equally gifted with the two latter perfections; but differ from the preceding in having a greater swell behind the ears, and a more evident degree of *throatiness*. The ewes of this flock are rather long in carcase. Their lambs, as well as those of the *Infantado* kind, are generally produced with a coarse hairy appearance, which is succeeded by wool of an excellent quality. The *Negretti* flock is composed of the largest and strongest sheep in all Spain*.

The Esserum sheep, when I saw them, had been eighteen months from Spain, and were in very good health. But two animals had been lost during the voyage, the inclemency of winter, and the abundant rains of spring, to which they had been exposed since their arrival in Denmark.

* The hairiness of the lambs here mentioned has been occasionally observed in England among the Negrettis also; and disappears when the animals are five or six months old. By some breeders it is esteemed a sure prognostic of an ultimate superior fleece. I believe that the Negretti was the only Merino stock brought to this country till last year, when the nation was enriched by the arrival of the Paulars, presented to his Majesty and the ministers by the Junta of Asturias.   T.

CHAPTER IV.

SAXONY.

UPPER Saxony is the country, into which, after Sweden, the Spanish race of sheep is of most antient introduction; and it is in Saxony that this naturalization has been marked with the completest success, and has produced the most advantageous results. The different indigenous breeds of that country, some of which produce valuable and others very coarse wool, have been equally improved by the Spanish sheep imported at two different periods, viz. in 1765 and 1768. The Elector of Saxony, wishing to repair the devastations occasioned in his dominions by a seven years' war, obtained from the King of Spain, in 1765, a hundred Merino rams and two hundred ewes, selected from the best flocks in Spain. Part of these animals were placed at the Electoral farm of Stolpe, six miles from Dresden, on the frontiers of Bohemia; and confided to the care of the Spanish Mayoral, who had brought them to Saxony.

Three other farms were also established at Rennersdorf, at Lohm, and at Hohenstein, which were principally devoted to the improvement of native breeds by the Spanish cross. Mr. Heinitz, at that time the Prussian minister of finance, was charged with the direction of these different establishments, under the inspection of the Electoral Chamber of Saxony.

It was perceived, after a lapse of ten years, that the pure Spanish race had preserved its properties, and that animals were to be found of several crosses by the Merino ram, which had acquired a fleece yielding to that of Spain neither in fineness, nor beauty. As soon as it was evident, from experience, that the Spanish ram could be easily reconciled to the climate, and the native breeds much improved by a series of crosses, attention was paid to the general improvement of flocks, after consigning to castration such animals as appeared defective. In 1776, sheep were disposed of; but as the best pursuits are always sure to meet with opposition, the sale was attended with difficulties so great, that the Saxon government obliged those, who occupied land under the Elector, to buy a certain number of the Spanish breed.

But the farmers were soon convinced as to their real interest, and the Electoral institutions not being able to supply the demands, which every day increased, the Prince imported from

Spain, in 1778, a hundred rams and two hundred ewes, of which a part were sold at prime cost, including the expences of the voyage.

It was thought that the Stolpe flock, composed entirely of the pure breed, should be increased, in order that the propagation of so valuable a race might be rendered more certain and more speedy. This augmentation has taken place, not only at Stolpe, but at the three other farms already mentioned; so that the number of pure Merino sheep, now belonging to the Elector, amounts to 3,400. Five hundred of these animals, which are annually sold by auction, prove insufficient for the demands made by the breeders, though it is also easy to produce these sheep in considerable numbers from individuals. This fact gives us reason to believe that the sheep at Rambouillet will be sought for with eagerness, and that the annual sale, at present so lucrative to the French government, will continue, for a certain number of years, to be of great benefit.

Circumstances did not permit me to visit the farm at Stolpe, or any of the other three, while I remained in Saxony; but I know from the accounts which have been conveyed to me on the part of those, who have seen the flocks

there, that the Merino sheep, naturalized in
Saxony during more than thirty-five years, have
preserved their original perfections. The wool
produced by these sheep differs in no respect
from the best in Spain, as may be easily perceived on examination of the specimens given
to me. This fact was confirmed in 1778 by a
comparison of the animals, which then arrived
from Spain, with the descendants of those imported in 1765; and the samples, preserved in
the establishment during twenty-three years at
least, allow no further doubt on the point.

I examined several flocks belonging to private
persons in Saxony; and I found that the pure
race, as well as the mixed breed when deeply
crossed, were carrying wool of the first quality.

The Merinos in Saxony are in general less,
and produce shorter wool, than in Spain. Some
of the former are, nevertheless, found, which
surpass the latter in these respects. The difference depends on the quality and quantity of
food, allotted to the animals on different farms.

I must confess that I have seen sheep, which
had palpably degenerated; but this was the result of bad selection, want of care, insufficiency
as well as improper quality of food, and un-
wholesomeness of the sheep-houses, where, ac-

cording to the custom of the country, the litter and dung are left throughout the whole year.

The Saxon government, which had devoted particular attention to the improvement of sheep, having ascertained by experience that the Merino flocks would degenerate if neglected, now applied itself to the instruction of farmers, partly by establishing schools for shepherds, and partly by distributing profusely works likely to be proper guides for those who resided in the country, as to the treatment of these valuable animals. The government wisely thought that it was alike its duty and its interest to assist the pursuits of agriculture; a measure which should ever be resorted to, when individuals have neither the means, nor the degree of intelligence, nor the spirit necessary in any novel and difficult undertaking.

Saxony has been amply repaid for the care, and indemnified for the advance thus made on the part of her government, by the immense advantages resulting therefrom to that country. The happy effects of this encouragement to agriculture have extended to other states of Germany, where the princes, wiser and more just than certain other potentates of Europe, generally think it a duty to watch over the internal administration of their dominions. A large portion of the taxes is often employed by them towards its proper purpose, that is to say, towards objects of instruction and public prosperity. The useful arts too, and above all, Agriculture, have made great progress for several years in various parts of Germany.

In Saxony, sheep-breeding is, at present, the most lucrative pursuit of the farmer, when conducted on the system of adopting Merino blood; for, exclusive of the profits which ordinary stock would yield, he gains considerably by the sale of these improved animals, and by the superior value of the wool, which often sells for thrice the return of a common kind.

The manufacturers, finding in their own country, wool in quantity and quality sufficient for their purpose, are no longer obliged to import this article from Spain; and are thereby released from the disadvantages attendant on the chances of commerce.

Saxony, which rears about 1,600,000 sheep of all kinds, can boast at this moment of 90,000 partly of the pure Merino race, and partly improved by crosses of that breed. It is calculated that these animals yield, one with another, 242 lbs. per hundred; which wool is sold at 2 ¼ francs per lb. Thus, calculating each sheep to produce about 2 ½ lbs., the 90,000 of improved stock, which this country possesses, will yield 225,000 lbs. of wool, and consequently return 562,500 francs. It should be observed too, that I have calculated the wool at a lower price than it often brings.

These sheep not only produce the quantity of wool necessary for the manufactures of the country; but furnish besides, a surplus equal to the interior consumption, which is sold at the Leipzig fairs, and goes thence to supply the fabrics of Aix-la-Chapelle, Holland, Flanders, Prussia, &c.*

CHAPTER V.

THE PRUSSIAN STATES.

FREDERICK the Second, who was not less illustrious for the wisdom of his administration, than for military renown, caused great im-

* A considerable quantity of it formerly came into this country, and some of it still finds its way. Our manufacturers esteem it highly; and I have specimens of Saxon-grown Merino wool, in all respects equal to the generality of Spanish piles.   T.

provement in the agriculture of Prussia. It
will be seen in the Memoirs of this prince by
Count von Hertzberg, his minister, that he
devoted to the above purpose the sum of forty
millions of dollars* between the years 1763 and
1778. Many districts of Prussia which were,
before that period, covered with water, or lying
waste, now produce abundant crops. It was
by the encouragement thus afforded throughout the reign of Frederick, and continued by
his successors, that Prussia, a country of natural sterility, has considerably increased its
produce, its population, &c. It is unfortunate
that these striking results are in opposition to
the theory of many persons, who pretend that
agriculture can prosper, and will attain perfection in France, without the assistance
of government. Would these men deprive
Bonaparte of a renown more real and more
durable than that, which he has obtained by his
martial exploits?

Frederick imported from Spain, in 1786, a
hundred rams and two hundred ewes, destined
to improve the indigenous breeds. Part of

* A Prussian dollar is about three shillings and nine
pence sterling. T.

these animals were placed at Stansdorf, near Berlin, and were lost by various disorders. The sheep, which were sent to different farms in the country, degenerated, in a great degree from the negligence of those, in whose possession they were placed. I have, nevertheless, seen flocks, derived from the original stock, which yielded wool as fine as their imported progenitors.

Since that period, several Prussian breeders, encouraged by the example of Saxony, have bought Saxo-Merino sheep, and have successfully turned their attention to the subject; so that there are, at present, in the Prussian states, flocks, which are entirely composed of animals yielding fine wool. Government has encouraged this propensity by instituting schools for shepherds; a sure way of disseminating good methods, and causing them to be adopted. Mr. Fink, a celebrated German agriculturist, is entrusted with the direction of the school established two years ago at Petersberg, in the province of Magdeburgh. Twelve shepherds are there annually admitted to learn the practice of their art.

Mr. Fink began the improvement of his flocks in 1756, by introducing the Silesian breed, re-

D

markable for the fineness of its wool. In 1768 he purchased some Saxo-Merino sheep, and in 1778 he imported a certain number of Merinos from Spain itself.

It is by the use of these foreign animals that Mr. Fink has succeeded in imparting the highest degree of perfection to the native breeds of Prussia, the wool of which was extremely coarse. This interesting agriculturist, whom I saw at his country residence, about three leagues from Halle, was kind enough to shew me in detail his beautiful and numerous flocks. The sheep are less than the Merinos found in Spain; but by no means inferior to them as to perfections of fleece. Before the improvement had taken place, the native breeds produced wool which sold from eleven to eighteen dollars per quintal of 100lbs. Mr. Fink sells that improved by the use of Spanish rams for sixty to eighty-five dollars per quintal. He disposes of about three-hundred sheep annually.

As the modes of treatment pursued by Mr. Fink in this part of the country, by the Count von Magnis in Silesia, and by those entrusted with the care of the Electoral farms in Saxony, offer to notice interesting peculiarities, remarkable differences, and results, the knowledge of which may be useful to persons, who devote

their attention to the improvement of sheep, I
shall take notice of these various methods in the
second part of this work.

The Count von Magnis, who has considerable estates at Eckersdorf in Silesia, to which
he retired in 1786, possesses at present 9,100
sheep, which he has improved by the breeds of
Hungary and Spain. When this zealous patriot abandoned the court, and devoted his attention to rural pursuits, he found on his estates
3000 sheep, the annual return of which was
rated at 1200 dollars. He has not only increased his flocks; but he has brought them to such
perfection, that the annual return is now considerably more than twenty-fold.

The Count von Magnis, who had at first endeavoured to improve the small breed of Silesia
by the large Hungarian race, has for some time
applied himself to the establishment of stock, in
which a good-sized carcase is furnished with
abundance of fine wool. With this view he
bought the handsomest Merino sheep which he
could procure in various parts of Germany,
often paying above a thousand francs for a single ram. Though the Count has not, even by
these purchases, yet reached the summit of perfection, it must be owned that he has travelled

as far towards it as he could, when the period is considered, at which he began his experiments. The wool yielded by the larger part of his flocks, will bear comparison with the best in Spain; and the animals surpass, both in size and symmetry, the most admired flocks of the Continent.

The Count von Magnis sells wool to the annual amount of 80,000 francs; which sum, added to the increment of his flock disposed of, makes a total of 105,000 francs, or 26,250 dollars. The price of his rams is seventy-two francs, and of his ewes thirty-six francs each.

This enlightened agriculturist, who manages his farms with a regularity and intelligence above all praise, has entirely abandoned the practice pursued by his neighbours. He has surmounted the obstacles, which an ungrateful soil opposed to him; and he has succeeded in supporting nine thousand sheep in a country, where the rigour of the climate prevents all pasturage during six months of the year.

The success, which has attended fine-woolled sheep in the Prussian states, and the attention bestowed by an enlightened government on this source of rural industry and wealth, afford reason to suppose that the number of pure Merinos, and of animals improved by that race, will continue to increase, and in process of time supply the growing manufactures of cloth. Not long ago Mr. Fink was sent to Spain for the purpose of obtaining a thousand Merino sheep, destined to this great national object.

CHAPTER VI.

THE AUSTRIAN STATES, AND OTHER PARTS OF GERMANY.

THE war, which desolated Europe at the time I travelled through the North, did not permit me to pass the frontiers of the Austrian states. Hence the accounts which I shall give of the Merino flocks in that country will be confined.

The Empress Maria Theresa imported from Spain in 1775 three hundred Merinos, which were placed at Mereopail, an imperial farm in Hungary, where a school for farmers was instituted. From this seminary, instructions were issued as to the treatment of fine-woolled sheep, and the modes by which perfection might be obtained through them. Shepherds were even sent to some individuals, who applied for them.

If political causes, which I shall explain in the second part of this work, had not opposed the progress of agriculture in Austria, the number of fine-woolled sheep would at this time have been considerable, when we take into consideration the expenditure of government, and several rich land-owners towards the improvement of the native breeds. Their efforts have been, however, impeded, and though the experiment may be said to have been crowned with feeble success, there still exist in the Austrian states, particularly in Bohemia and Hungary,* many flocks of the pure Merino breed and its crosses; the example of Saxony and Silesia having awakened attention.

Austria has procured, not long ago, two flocks of Merinos from Spain, which were embarked at Alicant for Trieste. The first importation, which arrived about twenty years since, was composed of three or four hundred animals, and the second of four to five hundred. An

* Among the various specimens of Merino wool from different countries in my collection, one from Hungary is certainly among the best, and surpasses many of the Rafino piles, imported from Spain.   T.

agent of the Austrian government is in Spain at the moment that I write this account, whose object is to obtain eight or nine hundred sheep.

The improvement of fine wool by using Merino blood is pursued in the Margraviates of Anspach and Bayreuth, by the institution of schools for shepherds, and by the results of an importation made in 1788, which consisted of forty rams and some ewes, principally of the Saxo-Merino, and Roussillon breed. In 1790, a second importation was made from Spain and Roussillon, which was placed at Rolenhof, and which has so effectually tended to the general improvement of sheep through the country, that there are at present but few individuals, whose flocks do not at least exhibit some symptoms of approach towards perfection.

There were in 1797 at the Rolenhof farm, 425 sheep of the pure race; and 8,141 of the improved breeds in the principalities of Anspach and Bayreuth.

The sale of animals as pure, which possessed only one, two, or three crosses of the Merino race, having counteracted the efforts of several breeders, and opposed a serious obstacle to the progress towards perfection ; Government endeavoured to remedy this abuse by publishing instructions, and pointing out where sheep were to be procured, which were indisputably genuine and well adapted to secure the growth of superfine wool. The flock-masters too found it difficult to procure a sale for their wool, partly on account of wanting customers for a home-grown article hitherto unknown, and partly because when they did sell, the manufacturers depreciated the commodity, that they might pay for it a price much inferior to its real value. The government, perceiving this, has secured an open and advantageous market to the wool-grower, thereby aiding an important branch of rural economy.

The Duke of Wirtemberg, who had occupied himself with success in agricultural improvements, resolved in 1786 to import fine-woolled sheep into his dominions. At this period he obtained from Spain and Roussillon 100 sheep; and he dispatched two shepherds to be placed under the care of Daubenton at Montbar, for the purpose of being practically instructed in their pursuit. During the next year he sent one of his counsellors thither, who, after having observed the methods pursued at the Montbar farm, went with the two shepherds into Spain, and there purchased a second flock, consisting of forty Merino and twenty-five Roussillon sheep. Of this flock, nine were lost on the journey through France, Savoy, and Switzerland. On the arrival of the rest, they were placed at Justingen, where they have gradually increased to the number of five hundred. The sale which has in consequence taken place annually, has contributed, in a material degree, to the general improvement of the native breeds.

The thirty-two Merino sheep, which were presented by that brave and skilful Commander *Moreau* to the Strasburgh agricultural society, were derived from the Duke of Wirtemberg's flock. These sheep, which were gratuitously offered to the French General by the Provisional States of the Duchy, after the conclusion of the armistice, are placed near Strasburgh in the bailiwick of Sulz, under the direction of a society, not less to be admired for its knowledge than for its attention to the most useful pursuits.

Merino sheep have been introduced into many other parts of Germany, where they have equally prospered. Baron von Molk, whose possessions are in Mecklenburgh, has obtained a considerable number of them, which he maintains on his farms there with great success.

The agricultural society of Zell supports, at its own expence, a beautiful flock of sheep, improved by the Merino cross, till its wool has attained the utmost degree of fineness.

Several flocks in Hanover, Brunswick, the Palatinate, Suabia, Baden, &c. have been improved by the introduction of Merinos. These sheep were imported into the Duchy of Brunswick during the year 1783, and into the Margraviate of Baden during 1788. The Margrave sent at that time Mr. Volz to purchase the excellent flock which is now at Pfozzheim.

CHAPTER VIII.

FRANCE.

It has long been suggested that France would derive great advantages, both with reference to agriculture and commerce, by improving the breeds of her wool-bearing animals.

Colbert was the first, who turned his attention to this important branch of national economy. This minister, as zealous for the prosperity of the public, as he was indifferent about his private fortune, had formed a design of improving the French flocks by importing, from Spain or England, sheep more perfect than those to be found at that period in his native country. The views of Colbert were useful, and founded on well digested consideration; but they were also new, and it, therefore, follows of course that they met with opposition.

The project had been alluded to at different periods, as various extant works tend to prove. It is stated in the *Nouvelle Maison Rustique*, (Edition 1721, page 317.) that Spanish sheep had already been imported at various times, with a view to the amelioration of the French breeds. The author of this publication expresses himself as follows:

" We have many Spanish sheep in different districts of France, and they succeed better, as well as multiply more extensively, than the English breeds; consequently it is easy to spread the race throughout the kingdom. By such a system, twice or thrice the profit now yielded by our common flocks, will certainly be procured, whether we take into consideration the strength and excellence of the rams and lambs, or the fecundity and milk of the ewes, or further the quantity and quality of the wool; or lastly the superior value of the skins."

This passage is the only mention made by French writers of a Merino flock at so early a date; and there is no trace of any breed improved by that cross, as the author elsewhere pretends.

Monsieur de Perce made some experiments about the middle of the last century, of which an account was published in the *Gazette de France* dated 30th December, 1752. These were attended with happy results, or at least had the effect of exciting public attention, and preparing the means, which have led the French nation forward to their present point of improvement.

From that time the amelioration of the fleece was pursued with renovated zeal and lively interest, and the goal would doubtless soon have been reached, if, instead of forming systems, nature had been consulted as to the causes by which superfine wool is produced, and the means, by which the quadrupeds, supplying it, might be preserved in their purity. " All those (observes Carlier in his Memoirs of Wool, 1755, page 99) who have written on the methods of bringing our wool to perfection, have differed from each other in their line of march. Some, hurried away by the ardour of their zeal, and the force of their imagination, have signalized their efforts by speculations rather far-fetched than solid, which, while apparently useful, were practically unprofitable. Others at once viewing the object in a less favourable light, have constantly denied the possibility that wool can be grown in France, equal to the piles of England and of Spain."

There has, however, appeared in France a skilful and judicious observer, who has rapidly caused the improvement in the flocks of that nation to pass from infancy to a state, which may be called adult. To him is attributable the basis, which imparts to his successors the hope of reaching perfection; a hope pursued by him to the utmost, during a long and laborious life.

The few, who are engaged in objects of great public utility, will foresee that I allude to Daubenton, who has devoted his mind, with a success equalled only by his perseverance, to the study of a race so important to our agriculture, to our subsistence, to our clothing, and to a multiplicity of arts, connected with our innumerable wants. I cannot, then, render to this illustrious and learned man homage more pure and more deserved, than by borrowing his own words from his excellent treatise intitled "Instructions to Shepherds, and Owners of Flocks." The passage which I shall quote, will present to the reader a succinct account of the progress, which has been made in France towards the attainment of perfection in the breeds of sheep.

"In 1766," says Daubenton, "Daniel Charles Trudaine, intendant of finance, which office embraced the department of commerce, foresaw that the Spaniards would refuse to furnish us wool, as soon as they had established manufactures extensive enough to employ all that of their country. Trudaine felt how severe a blow this would be to our commerce, because we could no longer proceed in the fabrick of fine cloths. He, therefore, endeavoured to devise some mode of preventing this injury to our national prosperity, and at the same time of emancipating France from a kind of tribute, amounting to several millions, which had annually passed out of the country into Spain, as a return for the wool thence derived. This mode was singular; it had for its object to grow in France wool as fine as that imported from Spain, and as suitable to our superfine manufactures.

"Monsieur Trudaine and his son did me the honour to consult me on this subject in 1776, and I informed them that the observations, which I had for a considerable time made on mingling the breeds of domestic animals, induced me to think that, by a proper selection of stock, sheep might be obtained producing wool both finer and longer than France at that time possessed. In conformity to this report, Mr. Trudaine proposed that I should enter on a course of experiments, which I did with a greater conviction of ultimate success, from the circumstance of the French climate appearing to me more favourable for sheep than that of either Spain or England; because there is not in France so great a degree of heat as in Spain, nor so many fogs as in England.

"The Messieurs Trudaine obtained from Monsieur de Laverdy, then comptroller-general of finance, all that was necessary towards my experiments, and government successively procured rams and ewes from Roussillon, Flanders, England, Morocco, Thibet and Spain. I placed all these animals on the farm, which I had established at Bourgogne near Montbar, in a district somewhat mountainous, and consequently favourable to the production of fine

4

wool, which was my leading object. I erected no cots; but kept the animals in the open air throughout the year, without shelter of any description; which experiment was crowned with complete success, and an account of it stated in 1769 to a public assembly at the Academy.

"I put very fine-woolled rams with ewes, which carried as much hair as wool, in order to judge by these extremes, how far the effect of the ram upon the fleece predominated over that of the ewe. I was very much surprised to see the offspring of such an alliance carry superfine wool; and this great amelioration inspired me the more with hopes of ultimate success in my enterprise, because it had been effected by a Roussillon ram, as I at that time had no Spanish sheep.

"In 1776, I received both rams and ewes from Spain, at which time I had seven distinct breeds of sheep, including that of Auxois, in which district my farm is situated. I have kept all these to the present time in their original purity, that I might see to which the preference was due; and I have also mingled these seven breeds among each other, that I might, by a variety of crosses, ascertain the degree, in which they influenced each other, with respect to improvement of fleece.

" By acting on these observations, experimentally followed with great caution till no doubt remained, I have brought all the breeds on my farm to a perfection of wool, which equals the piles of Spain, and without procuring any new supply of rams, either from that country or from Roussillon."

Daubenton, after being thus convinced of his point, caused cloths to be manufactured from his wool in 1783, when he sent 832 lbs., which had been washed on the backs of the animals, to Chateau-du-Parc, near Chateauroux in Berry. The manufacturer, after having converted this wool into cloths of various colours, engaged to pay for it the highest price allowed for imported Spanish. The following year another experiment was made, when the cloths produced were more flexible, and as soft as those from the first Spanish piles. It was even observed, that this improved French wool had a greater degree of strength than the Spanish, though still perfectly pliable. Cloths from Daubenton's wool were also manufactured at Van-Robais, Abbeville, Decretot, and Louviers. They exhibited the same perfections as others.

from superfine Spanish wool, which were made at the same places for the express purpose of comparison. Lastly, the fabrick at Gobelins was equally successful, and the article took a fine scarlet dye.

The sheep of the Merino race, which Daubenton received for the improvement of his flocks, were part of the flock originally obtained from Spain, by Trudaine, viz. the first which came to France, composed of two-hundred animals, which were divided among individuals in different provinces. It appears that the flocks, deduced from them, have degenerated, or have been entirely lost, excepting that belonging to Daubenton, which passed into the possession of Monsieur Thévenin, and another belonging to Mr. Barbencois, who originally received forty of the importation. The descendants of the latter still exist at Villegongis, in the department of the Indre. Thévenin, who is a man worthy to succeed Daubenton in such a property, has removed the flock to Tanlay in the district of Tonnere, where he proposes to enter on a course of experiments, the result of which will doubtless be most useful.

The wool of the Villegongis flock was this year presented to the agricultural society in the department of the Seine, and equalled in quality the first Spanish piles. This fact, added to a great number of others in my work, proves that the Merino sheep are reared with facility, and that they constantly produce superfine wool, though transported to a soil and climate differing from those of Spain.

It is true that this breed has not much increased in the department of the Indre, which Monsieur Barbencois jun. accounts for by a reference to the wretched treatment, which has attended them since their arrival in that country.

The farmers are, it appears, obstinately determined to keep as many Spanish sheep upon their land as they used to have of the native breed, though the latter were much smaller, and consequently did not require so large a portion of food. This mistaken conduct causes the new sheep to find, neither in the meadows nor the cots, a sufficient degree of nourishment, and hence arise frequent maladies, which have brought disgrace on the breed, and prevented its adoption in the neighbourhood. These animals, subjected to treatment so ruinous, are diminished in size, and are now, in this respect, only on an equality with the native breeds.

The farmers, who have thus suffered by an extensive mortality in their flocks, have replaced the original native race in lieu of the Spanish, and thus a mixture of the two has gradually taken place, to the inevitable debasement of the Merino. It has been also imagined that this breed was not adapted to the soil and climate; and that it could not be preserved in its pristine purity. The prejudice too, which attaches ideas of beauty and good properties to certain forms, has contributed to the depreciation of fine wool, and has prevented its sale on terms equally advantageous with that of the country. Such are the causes, which have produced a neglect of the Merinos in this part of France, and a gradual decrease of the breed.

These animals, which it would have been easy with care and attention to have spread through the whole department of the Indre in twenty-four years, are at this moment to be found only in limited numbers, and have but feebly contributed to the improvement of wool in the district.

The experiments and writings of Daubenton having, however, proved to government and several enlightened individuals, that it was easy to breed, rear, and preserve the Spanish race in

France, that the farms would be thereby bene-
fited materially, and that the manufactures
would be supplied with a most desirable resource, a considerable number of Merinos were
procured from Spain. Mr. Dangevillier, then
governor of Rambouillet, among others, requested of the Spanish government to be supplied with a Merino flock.

The king gave orders that a selection should
be made from the flocks of highest repute in
Spain; and in 1786, three-hundred and sixty-se-
ven rams and ewes were sent under the direction
of a mayoral, and three other Spanish shepherds.
These animals were driven, by short and gentle
journeys, to Rambouillet, where they arrived,
after having passed the winter upon lands in the
neighbourhood of Bourdeaux. About sixty of
them perished in the whole course of their passage.

The flock had hardly been fixed five weeks at
its destination, when it was perceived that many
of the sheep were infected by the *claveau* *.

* The *claveau*, *clavelée*, or sheep-pox, is a truly formidable disease, now happily unknown among the flocks of
Britain, and long may it remain so! Mr. Lawrence, however, is of opinion, that it has appeared in this country
at various times, but does not conceive that it was ever so

This malady would have caused great ravages, had not the necessary precautions been immediately taken. Even in spite of these, thirty-five ewes and sixty lambs were lost. Since that time, no attack of this kind has occurred, care having been taken to prevent all communication with neighbouring flocks, and all use of adjoining commons.

The Spanish shepherds attended to them for six months jointly with the French; and after the departure of the former, the bailiff of the farm devoted a considerable portion of attention to them. This was Mr. Bourgeois, a man who unites with very extended agricultural knowledge, a taste and talent, admirably suited to experimental improvements.

The present establishment at Rambouillet was, before the revolution, a farm dependent on the Chateau-Royal of that place. It is to be

fatal here as abroad. He adds, with respect to this malady, that " the thanks of the country are justly due to the patriotic cautions of Sir Joseph Banks, on the possibility of importation with foreign sheep." It appears to me that a re-print of these and any additional cautions would be particularly desirable at the present period, when a large flock of sheep from Spain is about to be distributed through the country.   T.

lamented that Tessier, who had commenced a series of experiments on this farm, was checked in the midst of his useful labours, and that circumstances did not permit him to attain results, which would, without doubt, have had a great influence on the improvement of our agriculture. At present there are kept on these lands a score of cattle, which are of the original wild race, thirty cows and heifers derived from the union of a hornless bull with Swiss cows, a few stallions, and the Merino flock. The friends of agriculture see with regret that a farm, sufficiently large for experiments of every kind upon no mean scale, is entirely devoted to this inconsiderable portion of animals.

When the original Merino flock arrived at Rambouillet, it was composed of sheep possessing uncommon beauty, and such as had been hitherto unknown in France among all the flocks which had at different periods been procured from Spain; for these having been obtained at many different places, considerably distant from each other, they were distinguished by striking local differences of appearance. They exhibited a variety, which was disagreeable to the eye, although of no consequence with regard to their quality. These characteristic distinctions arise, in some degree, from the repeated alliances of those animals, in which they first appeared; and the result is a race, probably differing materially from any of the animals, which formed the original stock, but yielding in no respect to the best of them, whether we look to size, to form, to hardiness; or to fineness, length, softness, strength, and abundance of wool.

The Rambouillet flock gradually increased, till it was deemed numerous enough to allow the sale of sheep annually, for the purpose of distributing them through the country. At first, indeed, several rams and ewes were given to individuals; but as soon as it was perceived that they were considered of little value by the farmers, because they were distributed as presents, an annual sale was determined on. The provincial administrations, then established, applied for them, and had a preference.

The flock at Rambouillet was, from the commencement of the revolution, placed under the care of an agricultural board, which preserved this precious depôt from the destruction so often hanging over it. The members of this board, zealous friends of the public good, were Messieurs Bertholet, l'Heritier, Cels, Vilmorin,

Dubois, Gilbert, Huzard, Parmentier, and Rou-
gien-Labergerie, to whom Tessier was afterwards added.

I shall here present to the reader a table describing the general produce of wool, and sale
of sheep from the Rambouillet flock between
the years 1796 and 1801, both inclusive; a period at which the Merino breed made rapid
progress, and the agriculturists of France began
properly to appreciate the resources, afforded
by this race.

Table of the Rambouillet Flock.

Year.	Average weight of the fleece in the yolk.	Average Price of the fleece.	Average price of sheep sold.		Highest price of sheep sold.
1796	6 lbs. 9 oz.	5 francs	Rams	71 francs	200 fr.
			Ewes	107 do.	
1797	8 lbs.	12 francs	Rams	64 do.	120 fr.
			Ewes	80 do.	
1798	7 lbs.	10 fr. 50 cent. . .	Rams	60 do.	150 fr.
			Ewes	78 do.	
1799	8 lbs.	15 fr. 78 cent. . .	Rams	80 do.	120 fr.
			Ewes	68 do.	
1800	8 lbs.	24 francs	Rams	331 do.	510 fr.
			Ewes	209 do.	
1801	9 lbs. 1 oz.	27 fr. 95 cent. . .	Rams	412 do.	630 fr.
			Ewes	236 do.	

Those acquainted with the wools of Spain
and France, will perceive, by casting an eye
over the above table, that the Merino fleeces
obtained a price far superior to that of any

native breed, however highly esteemed; that these prices increased, and in the two last years reached the sum given, in the common routine of business, for the superfine piles of Spain.

It should also be observed, that the merchants and tradesmen have constantly been forming coalitions, to depreciate these wools, and to prevent the possibility of a sale at their real value, besides which, various other circumstances operated against them.

These observations prove, at all events, that the wool grown at Rambouillet is as good as, and perhaps better than, the imported piles of Spain, and consequently that the adoption of this race must secure to the breeder a profit far superior to any, derived from the indigenous breeds of France. These animals not only produce wool so much finer, but they yield a greater quantity of it, as may be seen on reference to the table which I have laid before my readers; for the average weight of a Merino fleece is 7 to 8 lbs. in the yolk, and there are some particular animals, which even yield 12 lbs.; nay, one has this year produced 16 lbs., and the same weight has been obtained in some private flocks; while the common sheep, in the neighbourhood of Rambouillet, and in the greater part of France, produce no more than 3 or 4 lbs. each, generally sold for about half a franc per lb. which gives the pure Merino an astonishing right of preference, and the mixed breed a great one, in proportion to the number of crosses.

These advantages are obvious. It is true that they will not always exist in the same degree, because the extension of the Merino race must cause them to diminish. Still, however, before that can take place, a sufficient time will elapse to reward, both by the sale of wool and stock, those who first devote their attention to this great national pursuit.

The price of Merino sheep has materially risen at Rambouillet, in the course of the two last years. It has even reached a height which appears extraordinary in a country where we are not accustomed, as in England, to stake considerable sums on the acquisition of animals calculated for our purposes. There are many owners of pure Merino flocks, who will not, at present, part with a single sheep for less than 150 to 250 francs. The applications, which are made on every side, justify the belief that these prices will, for a considerable time, be maintained. They may even be expected to increase, as the prejudices, still existing in the minds of many, die away, and make room for better calculations of profit.

I must here make mention of a deception introduced into this species of commerce, against which buyers ought to be on their guard. Many breeders sell sheep as being of the fourth or fifth cross, when they are only of the second or third, or even the first, and again deeply crossed sheep, under the false appellation of *pure*. These frauds may essentially retard the amelioration of our French breeds, as has been the case in other countries. Imperfect animals, however they may be apparently gifted with the requisite qualities, cannot impart perfection, when put to sheep of any common breed, or even to the purest Merinos*.

* The same deceptions are not unfrequently practised in this country; and cannot be too severely condemned. The breeder, who, in beginning his experiment, relies on the purity of the ram, which he has purchased, is, perhaps, using a Merino-Ryland of the second cross. The consequence is, that the offspring do not produce wool of the quality, which he was led to expect; the price offered for it is not satisfactory, even if quite as much as it is worth, and the breed sinks into unmerited disrepute, not only in his estimation, but, from his report, among his neighbours. Every person, who thus imposes upon indivi-

The experiments of Daubenton sufficiently proved that the superfine wools, grown in France, were capable of undergoing every process, to which those of Spain are subjected, and that the cloths produced from them, were, in all respects, as good. Since his time, these experiments have been repeated, as well with the produce of the Rambouillet flock, as with others, and always with the same success.

In 1799, the agricultural society in the department of the Seine had several cloths manufactured from the Rambouillet wool, and from that of a mixed breed, kept at the ci-devant menagerie of Versailles, occupied by the Abbé Sieyès. I was appointed one of the committee to superintend the manufacture of these articles, and afterwards carefully examine them. All my colleagues, and indeed every person who saw them, agreed with me in a declaration

duals, and through them the public, ought to be publicly exposed; in order that those, who wish to give this race a trial, but are not completely acquainted with its distinguishing features, may not be the dupes of this narrow-minded traffic. He, who would really found a good Spanish flock, must, at the outset, resolutely put his hand into his purse, and pay his money at a market, where purity is indisputable.   T.

that they were equal, in all respects, to the cloths made from superfine Spanish wool.

Government also tried other experiments with the same view, and the following report was made by their two commissioners, Messrs. Tessier and Huzard, to the Institute in the year 1800.

" We have caused the pure Merino wool, and that of the mixed breeds, bought last year, to be manufactured. Messrs. Roi and Roui, Sedan, Decretot and Delarue, De Louviers, &c. have willingly undertaken these interesting experiments. It is but justice to say that Messrs. Roi and Roui were the first, who zealously determined on the use of French-grown Merino wool. They, and all the rest, have not only furnished us with good cloths made from the pure wool, but from that of the mixed breeds, and the latter, being exposed to sale, have made a very great impression on the growers. We learn that several of them, in selling their wool, have stipulated for a return of a certain quantity in the shape of cloth, that they may wear the produce of their own farms."

It is proper that I should take notice of some experiments, which have been made at Rambouillet for the purpose of establishing all the advantages which may be derived from the Merino breed, whether relating to the wool, or carcase.

The following experiment was tried in 1797, on the recommendation of Monsieur Gilbert. A ewe, then eighteen months old, was left unshorn. In the year 1798, her fleece weighed 14 lbs. 10 oz. The filaments of this fleece were twice the usual length, and it does not appear that any weight of wool was lost; for there are but few ewes, which would, in thirty months, produce this quantity.

A second case, which was also shorn at the age of thirty months, in the year 1799, yielded a still greater fleece, weighing 21 lbs. which was of a staple eight inches long, and something will be fairly allowed for the six months in which she was influenced by her lamb. The female of the former year produced in each one following 6 lbs.; and the ewe just before mentioned has since yielded 9 lbs.; whence it results that these animals have afforded as much wool, as if they had been shorn annually.

In the year 1800, eight ewes were shorn, which carried fleeces of two years' growth, and yielded 16 to 20 lbs. of wool each.

From these different experiments it appears
that wool, suffered to remain on the animal for
two years, acquires a double length of staple,
preserves its fineness, and loses nothing in point
of quantity.

It has not been observed at Rambouillet, that
the sheep, subjected to this trial, suffered much
from heat, or that their health was in any
respect injured. It was merely remarked that
the lambs had more difficulty in sucking, from
the length of those locks, which completely enveloped the udders of the dams. The conjecture formed on these experiments, is that, by
leaving the fleece on the animal for two or
three years, the wool may be employed with
great advantage in various fabricks, particularly
in the manufacture of cassimirs, as was the
case with the produce just described; but in
order to establish, on a certain foundation, the
benefits or inconveniences, which would result
from such a measure, it would be necessary to
let a whole flock remain unshorn for two years.
It would be desirable to know whether a certain
number of animals would not, in this case, want
a larger quantity of food; whether they
would not be more frequently subject to disorders; whether difficulties would not occur in

driving them to their pastures; and finally, whether other inconveniences would not ensue, which are unperceived in trying the experiment on a small number. It is probable that the inconveniences would materially exceed the advantages.

Experiments were also tried at Rambouillet for the purpose of discovering whether it was better to shear the lambs at about six months old, or suffer the fleece to remain till they became two-tooth sheep. The latter yielded as great a quantity of wool as those, which were twice shorn. The article too was stronger and longer; and it brought a better price.

Such are the results presented to the Institute by Messrs. Tessier and Huzard.

Mr. Yvard, who unites with a profound knowledge of agriculture, the most ardent zeal for the progress of that art, has made some experiments, which appear to contradict those at Rambouillet. He thought that, in order to have the most positive results, it was not sufficient to let the wool grow upon lambs till they became two-tooth sheep, and then to compare the produce with that of others shorn twice, since the latter might be more feeble, and not so well supported with proper food. This inconvenience Mr. Yvard removed by taking four lambs from his flock, and shearing them longitudinally, as to one half of the body, at two different periods. The wool of the other half was left unshorn till they became two-tooth sheep.

A lamb, of which one half was shorn in 1800, yielded $11\frac{1}{4}$oz. of wool. The same portion of his body produced in 1801, when he became a two-tooth, 5lb. 11oz. making a total of lbs. 6 $6\frac{1}{4}$

The side not shorn in 1800 yielded in 1801 no more than - - - - 5 $2\frac{1}{4}$

Advantage of the double shearing
lb.   oz.
1   4

Another lamb similarly shorn, 1st clip $12\frac{1}{2}$oz. second 4lbs. $14\frac{1}{4}$oz. Total 5 11

Side not shorn the first year - - - 5 $10\frac{3}{4}$

0   $0\frac{1}{4}$

A third lamb as above, 1st clip $8\frac{3}{4}$oz. second 4lbs. $8\frac{1}{2}$oz. Total - - - 5 $1\frac{1}{4}$

Side not shorn the first year - - - 4 9

0   $8\frac{1}{4}$

A fourth lamb as above, 1st clip $8\frac{1}{2}$oz. second 3lbs. 10oz. Total - - - 4 $2\frac{1}{2}$

Side not shorn the first year - - - 3 $4\frac{1}{2}$

0 14

Total advantage of the two shearings over the single clip of the four animals - - - - - - - - - - - - lbs. 2 $10\frac{1}{2}$

The results of this, and other experiments, by Monsieur Yvard, tend to prove that it is advantageous to shear the lambs; and further, that a loss is sustained by preserving the fleece two years on the back of older animals.

It may be objected to his mode of trying the experiment, that the humours, which cause the growth of wool, have a tendency towards that part of the animal, from which the wool has been taken; and that, consequently, they accelerate the growth of one part of the fleece, at the expense of the other. This supposition, which does not seem to have any sound foundation, would in reality operate in favour of the experiment, because it implies that by the clip the growth of the wool is increased.

I presume that a practice, pursued by some good agriculturists in England, might be advantageously adopted. This consists in cutting the fleece to a small depth from the surface, and thus taking away all the outer part, to the amount of about half a pound or more, in an animal of moderate size. The growth of wool is thereby increased, and an advantage gained at the period of general shearing. Experience has proved that hair must be thus treated, in order to make it grow vigorously.

It was at first supposed that the Merino race would not be capable of producing, in France, as good wool as in Spain; because France did

not possess the same soil, or the same climate, as
that Southern part of Europe. It was also said,
that the fineness of the wool being an effect of
the journeys, no such quality could ever be obtained in France, because it was impossible that
the animals could be similarly treated. All
these arguments having been refuted by expe-
perience, certain persons have endeavoured to
depreciate the Merinos, by maintaining that
they were not of a nature to fatten, and that
the mutton was of a very bad quality.

This objection, specious, perhaps, in some
respects, is, nevertheless, completely annihilated
by facts. It is true that the mutton in Spain is
generally lean, tough, and of a disagreeable
taste; but I have observed, in my Treatise,
that the bad quality of the flesh in Spain arises
from the circumstance of none being consigned
to the butcher until they are no longer of use
for breeding, or begin to decay from old age.
The butchers then convey them to the commons in the vicinity of towns, where the herbage is so scanty, that they can scarcely find
enough to subsist upon. The Spanish shepherds keep very few wethers in their travelling
flocks, and castrate the males in general only
at an advanced age. Every breed, subjected

to such improper treatment, would produce mutton of bad quality. The coarse-woolled sheep of Spain do not supply mutton more delicate than that of the Merinos; and I have eaten Merino mutton in this country, as savoury as that of the French breeds.

What I have here advanced is confirmed by the facts, which Messrs. Tessier and Huzard recorded. The following is a part of their report to the Institute in 1799.

" The experiments which we have tried as to soiling Spanish sheep have never been properly detailed. All the animals thus treated became fat, and their flesh was at least as delicate as that of any other breed.

Having to oppose a prejudice, founded on the idea that Merino sheep were not easily brought to proof, it was necessary that we should enable ourselves to record positive facts, distinguished by that authenticity, which characterizes experiments, made under the eye of government. Accordingly, on the 18th of Ventose, * three wethers were set apart. They

* I have not just now any means of reference to the exploded French calendar, for the purpose of ascertaining this date; but as we are afterwards informed that the dura-

were of the same age, but of unequal weights, making together 243 lbs. Between the heaviest and lightest there was a difference of 13 lbs. They were at first fed on lucerne and bran, the latter of which was afterwards suppressed, when oats and barley were substituted for it. The animals were weighed about every fortnight, as well as the food destined for them during that period.

The result was, 1st, that on the 14 Prairial, when one of them was slaughtered, the total weight of the three was 326 lbs., an increase of 83 lbs.

2. That the difference at this time between the one which weighed most, and the one, which weighed least, was 8 lbs. 8 oz., making the difference 4 lbs. 8 oz. less than when they were first put up to feed.

3. That the one, which weighed most at last, was the one, which weighed neither most nor least at first.

4. That the one, which weighed least at last, was the one, which weighed least at first.

tion of this fattening experiment was eighty-six days, it is perhaps not of material consequence to know the exact period of the year.   T.

5. That the greatest increase of weight in all of them took place in the first twelve days.

6. That during a hot season there was a remission of increase in two only; after which, the weather being cool, the increase became greater.

7. That at two periods one continued to fatten, while two others lost in weight, and that the two, thus losing weight, were not both times the same.

8. That after the 15 Prairial, the two remaining lost weight more and more, though unequally.

Each sheep consumed, in eighty-six days, 281 lbs. of food, or 3 lbs. 4 oz. per day. The one, killed on the 14th Prairial, which then weighed the least of the three, returned as follows:

	lbs.
Carcase - - - - - -	51
Fleece - - - - - -	7¼
Loose fat - - - - -	5½
Head, feet, skin, and intestines	32¼
Blood - - - - - -	3¼
Total - - -	99¼

We must observe that, among the Merino

sheep before fattened, several had produced
more loose fat than the one, which furnished
occasion for the above details. Doubtless the
other two yielded more; but we could not then
dispose of them. However that might be, the
mutton was acknowledged to be excellent.

This experiment, among others, is a decisive
proof that Spanish sheep are capable of being
made fat; and that the objections to the quality
of the meat are groundless. The interest of job-
bers and butchers has principally given rise to
this false report; they wishing to obtain these
sheep, and the mixed breed derived from them,
at a low price. The prejudice is without foundation, and tends to retard the progress of improvement; hence we deem it important that
truth should be opposed to it."

It has appeared, in the course of this work,
that every government, which tried to introduce Merino sheep, has not been content with
obtaining the animals from Spain, and encouraging the propagation of them; but has
thought that all these measures would be in-
complete, unless instructions were properly disseminated. The Board of Agriculture, as well
as the Minister of the Interior, have been of
this opinion in France. Sensible that the pro-

pagation and success of the Merino breed would be in proportion to the knowledge of the farmers and shepherds, they issued a publication with respect to the treatment of sheep, which was drawn up by Gilbert, and founded a practical school for shepherds at Rambouillet.

This institution has proved of great service; for several shepherds have been sent from it into different departments; and such will be the case more and more when land-owners in easy circumstances turn their attention to rural economy. Seven or eight pupils are instructed at Rambouillet, and paid for by the departments; while individuals have the power of sending others by paying thirty-six francs per month.

In order to facilitate the sale of Merinos, and their produce of superfine wool, a fair was last year established at Rambouillet, which is intended to be held at the same time and place annually. It will doubtless provide individuals with an ultimate opportunity of selling their wool; but at present the buyers, who attend it, have entered into an agreement among each other to depreciate the article, and the first sale has, in consequence, not been favourable to the growers. The fair having been fixed too at a period so immediately after the shearing-season, the farmers have not time to send their wool. It would, therefore, be advisable to appoint this meeting a month later, and it would be still better if it were held at Paris, or in the environs; for though a great number of the pure Merino breed, as well as the mixed, are on farms at a distance from the capital, yet the breeders and graziers so constantly resort thither on a variety of business, that they would sell their wool, and not be burthened with any extraordinary charge.

A further efficacious assistance might also be rendered to the sale of superfine wool, by establishing in the environs of Paris a lavatory, similar to those of Spain, for the use of the wool-growers. Others might successively be stationed in those departments, which contained sheep enough to support an establishment of this nature.

The first lavatory should be made at the expence of government; for no individual would be willing to enter on such a speculation. The expence of constructing it would be trifling, and this might be repaid by a moderate charge for the use of it. After the article was thus washed, it might be fairly and easily offered for sale as Spanish wool, and would obtain its real value. The growers might themselves dispose of it to the manufacturers; or they might employ brokers to defeat the purposes of those who coalesce, when they treat directly with the growers.

There are in France two establishments, at which the Merino sheep are kept, exclusive of the one at Rambouillet. Government thought that it was not less important to propagate the breed in the South than in the North, and that flocks ought to be established at different points for the purpose of proving, by example, how far they would be equally successful in all parts of France. It became also necessary, for the purpose of encouraging this branch of industry, that the owners of flocks should have an easy opportunity of improving them.

With this view, a Merino lot, consisting of six rams and seventy ewes, was sent in 1797 to Pompadour in the department of the Corrèze. The establishment at this place had hitherto been destined to the care of horses and cattle. In 1800 there were still on the farms nineteen stallions, thirteen brood-mares and foals, two Tuscan asses, nine Limosin and eight Romagne cattle, with six of the wild race. By the arrival of the above Spanish flock, this institution acquired a new degree of utility, and the bad breeds of the neighbourhood insensibly disappeared.

The Pompadour flock is at present composed of two hundred and forty-nine sheep, which are under the care of a shepherd, instructed at the Rambouillet school.

The establishment at Perpignan was founded by Gilbert, who sent thither part of the sheep imported from Spain. He had judiciously foreseen that fewer obstacles would be there opposed to the adoption of this breed than in some other districts, and he there made more rapid progress; partly because the inhabitants of that country were already familiarized with the form proper for the Spanish race, and better knew their excellence than the farmers in other quarters, and partly because a flock of this nature was certain there to meet with many other local advantages.

Government, having adopted Gilbert's plan, gave orders for the necessary arrangements to be made at Perpignan, and a shepherd was sent thither from Rambouillet to take charge of the Merino flock. This consisted at first of three-hundred and sixty animals, and by breeding is now increased to seven-hundred and thirty-five.

It remains for me to speak of the mixed race at Alfort, in order that I may lay before the reader an account of all the public institutions for the improvement of sheep.

The flock, which is now at Alfort, was originally formed at Sceaux, and thence removed to Versailles. Sceaux was an experimental farm, established by the Republic in the year 2 (1793); but political reasons, or such as had the semblance of propriety, having caused it to be sold, the French Directory, struck with the justice of the observations contained in the report which the Institute presented on that occasion, and being convinced of the advantages to be derived from a national agricultural establishment, ordered the Merino flock to be removed to the menagerie at Versailles.

The numerous experiments in different branches of rural economy, which were begun at Sceaux, and continued at Versailles, have entirely ceased since 1799; the menagerie having been then awarded to Citizen Sieyes, in return for the important services, which he had rendered to the state.

It is to be regretted that government was not

sufficiently acquainted with the innumerable advantages, which the agriculture and industry of France would hereafter have derived from the experimental farm at Versailles. Had such been the case, some other recompence would doubtless have been bestowed on Citizen Sieyes; and the latter would have been anxious to apply for an equivalent object, had he properly studied those arguments founded upon public utility, which were presented to him, with this intention by the agricultural society.

" If agriculture be the first of the arts," (says the Institute in its report;) " if it cannot be brought to perfection without experiments; if these experiments require a national farm; if it be true, as an eminent practical writer asserts, that a government, which has not such an establishment, is in want of a most useful institution; if the farm of the Menagerie at Versailles be, at present, the only national estate adapted to such an object; and if it be, notwithstanding all this, without hesitation taken away, and converted to other purposes; where, we would ask, are there any public establishments, which will be respected? Under what circumstances will the interest of the public be able to triumph over the interest of an individual?"

The flock of the mixed breed, transferred from Versailles to Alfort in 1799, was formed with a view to experiments in crossing; thereby to ascertain the best mode of acquiring the variety of wools, so necessary in the manufacture of those cloths and stuffs, which form the principal branch of our commerce in the Levant.

This flock, which consisted of two hundred and fifty-four sheep and sixty-three lambs, was diminished by the loss of twenty-seven animals soon after its arrival at Alfort; and in 1800 the total was two hundred and eighty-six.

The *Valais* sheep, which surpass the French breeds in size, carry very coarse wool resembling goat's hair, which is unequal both in length and quality.

Those of *Bearn*, not so large as the former, yield wool as long, quite as coarse, and extremely full of *kemps*. Both these breeds produce a part of the wool shorter and finer than the rest.

The fleeces of the *Beauce* sheep are not of a staple so long as is grown by either of the above kinds. The wool is not so dry, so brittle, and so debased by the admixture of *kemps*. It is also more equal than that of the preceding breeds.

The *Boulogne* sheep are strong, and in countenance, as well as form, resemble some of the English breeds. They yield stout wool, which possesses a certain degree of elasticity, but is altogether of coarse quality.

The English race known by the appellation of *Lincolnshire*, is also at Alfort. The animals are eighteen to nineteen and a half inches high. The carcase is thick and short, the legs stout and long, the head square, the mouth and muzzle expanded. This breed is of a hardy constitution, and produces wool, which is soft, and possesses a little elasticity.

The *Sologne* race is one of the smallest in France. It is characterized by short legs of fine bone, a small head and ears, a smoky or black face, as are also the legs. The wool, which is short and curled, should be ranged at the head of our second-rate indigenous piles.

The *Roussillon* sheep are generally somewhat less in size than the Merinos, and their wool almost equals that of the latter in fineness. It differs, however, in being not so close, as it grows on the animal in detached locks. The filaments have a spiral appearance from one end to the other. The Roussillon sheep are remarkable for the yellow tinge of the fleece, which contains a large quantity of yolk.

By putting the ewes of these seven different breeds to Merino rams, a mixed race has been obtained, yielding a product more or less fine, and more or less resembling the Spaniard. The wools are stamped with a character, which indicates the maternal race of the animals producing them, and they have acquired a greater perfection by a repetition of crosses, now carried forward to the fourth generation. At this period they have attained the particular degree of fineness, which they are capable of reaching. Breeds, originally possessing wool of superior quality, would, by a combination with the Merino ram, sooner attain this degree. These mixed breeds are kept at the point of amelioration which they have reached, by alliances among themselves, without further recourse to the pure Spaniard.

If we breed from a Merino ram and a long-woolled ewe, the wool of the offspring will be longer than that of the father. On the contrary, it will be longer than that of the mother, if she were short-woolled. The length diminishes in

G

the first case, and increases in the second, proportionately to the distance of the animals from the first generation; so that the difference is hardly discernible after four crosses.

A sheep of the mixed breed will produce a greater quantity of wool when the fleece of the dam was close in its texture, and a smaller quantity if it was open.

The abundance of wool depends particularly on the size of the animal, as well as on the quantity and quality of the food assigned to it. No one must expect a great return in wool, unless he allows the sheep a sufficient portion of nutriment.

The sheep derived from a Spanish ram, and any native breed, gain or lose in size according to that of the dam, whether larger or smaller. Hence, those breeders, who want a large race of the mixed blood, should put to Merino rams ewes stronger than themselves; while those, who are attached to small sheep from the nature of the pasturage upon their farms, should select ewes inferior in size to the ram. The influence of the male being greater in generation than that of the female, it is important to choose large or small rams; or such as are gifted with any certain quality according to the kind of flock, which it is proposed to form.

The sheep from the Merino cross upon native ewes are, like every race, capable of acquiring new qualities as to fleece, carcase, form, &c. Perfection is only to be acquired by selecting, in every generation, those individual animals, which possess in the highest degree the quality required, and employing these only for procreation.

The seven breeds, which I have mentioned, are of robust constitution, with the exception of the Sologne, which is very subject to disorders. The animals derived from the Spanish cross on this race are more hardy; but they fatten less kindly than those obtained from the Boulogne, Lincolnshire, Beauce, and Valais breeds. The same inconvenience attaches to the Merino-Bearn, and Merino-Roussillon breeds.

Those which soonest improve as to fleece, are the Boulogne, the Lincolnshire, the Sologne, and particularly the Roussillon, because it most resembles the Merino. The Bearn is more susceptible of amelioration than the Valais, though these two breeds yield wool almost equally coarse.

I presume that those ewes, which carry

fleeces composed of very coarse filaments, and of others equally fine, but shorter than the former, and more capable of producing good wool-growers than those, which bear fleeces not quite as coarse, but uniformly so.

It would be a most important step in the progress of improvement, if we were to dive into the causes, which combine to render any race of sheep capable of more rapid and complete amelioration; but a long series of experiments must be tried before we can acquire notions proper to guide us in so deep an investigation. Professor Godine of the Veterinary College at Alfort, who was kind enough to communicate most of the facts here recorded, and who shewed me the flock at Alfort, confided to his care, is at this moment pursuing a train of experiments, the results of which will not be less useful to science than honourable to the enlightened individual, who is so zealously laborious to attain beneficial knowledge.

A school for shepherds has been established at Alfort, where twelve to fifteen young men, sent by the departments, and owners of flocks, are educated. The prefects of the departments give thirty-six francs per month for each pupil. This sum is for their bed and board, dress, books, and implements necessary in the course of instruction. Private individuals pay twenty-two francs more. The shepherds are boarded and lodged with the veterinary pupils, and subservient to the same discipline.

The instruction of these young shepherds is of a year's duration, which period is divided into three parts. In the first four months, they are instructed in the elementary anatomy of sheep, the principles of physiology, physic, &c. The four following months are employed in teaching them the natural history of the sheep, the means of improving breeds, the manner in which flocks should be treated as to food and care, the fold and cot, &c. During the last four months, lessons are given on the diseases of sheep, surgical operations, preparation of medicines, &c.

. It is proposed to introduce at Alfort all those French breeds, which bear a distinguishing character, for the purpose of forming a series of crosses by the Merino ram, and thereby to discover which is entitled to a preference with respect to wool, carcase, health, economy, &c. It would not be less useful to cross some of the foreign breeds, such as those of the Texel, Friesland, Moldavia, Padua, &c.

The immense advantages derivable from the flock at Rambouillet, are no longer hypothetical. Those, which we may expect from that at Alfort, though less considerable, will, perhaps, not be less real. It is to be regretted that such an establishment should be situated where it has not an acre of land belonging to it; and that animals destined to a course of experiments, which require particular vigilance and care, should be obliged to graze on hired pastures.

There exist in France, besides the national flocks at Rambouillet, Perpignan, Pompadour, and Alfort, a considerable number of pure Merinos, and of sheep improved to the highest degree. It appears, from the last returns delivered by the departments, that there are in France a million of these animals; and supposing these accounts to be exaggerated, it is, nevertheless, certain that the number is much increased within the last year. There can be no doubt too that a still more rapid increase will gradually take place, when we consider the eagerness, with which the Merinos are now sought after, the high prices which they bring, and the value attached to their wool in the market. We have reached a period, at which prejudice is beginning to subside, and a con-

siderable number of breeders, as well as manufacturers, are now enlightened with regard to
their real interest.

Having thus given an account of the Merino
breed and its present state in France, I shall
now speak of the measures, lately taken by
government, to insure the propagation of this
race, or rather rapidly to increase the number
of these animals.

The Directory had reserved to itself by a se-
cret article in the treaty of Bâle, the right of
bringing by purchase from Spain, during five
successive years, one hundred rams and one
thousand ewes per annum; but, for some time,
no attempt was made to obtain a stock, replete
with such advantages to French agriculture and
industry. At length Gilbert, whose only wish
was the public good, read in 1797 a Memoir
to the Institute on the state of the Rambouillet
flock, wherein he developed his ideas on this
subject, and pointed out the means, which he
thought best adapted to the attainment of the
stipulated importation, as well as most useful
towards the propagation of Merinos in France.
He stated, that government ought to bring these
sheep at its own expence, and by no means
cede any part of its right to individuals; adding,

that it would act in immediate contradiction to its own views, if it imported so considerable a number at once. I opposed Monsieur Gilbert's ideas in my treatise, which appeared soon after the Institute published an account of its proceedings. These were my words: " I am of opinion that if certain individuals would bring the sheep from Spain at their expence, whether intending to keep, or to dispose of them, government ought not to prevent it. However great might be the number sold by these importers, the improvement of our fine wool would suffer no abatement; for no one will employ his capital either in sending to Spain for the sheep, or in buying them after arrival at second-hand, unless he expects to make a profit thereby; and this profit he knows that he cannot expect unless he bestows upon the breed that care, which it requires. Hence, so far from this speculation being detrimental to the improvement already begun, it is evident that the reverse consequence will ensue, and that if the scheme be carried into execution, our manufactures will, in a few years, be independent of Spain."

The measures pointed out by Gilbert were adopted as far as related to the ejection of individual importation, and government, which decided on bringing the five-thousand five-hundred Merinos at its own expence, sent him into Spain for the purpose of selecting and purchasing. that number. Again we must lament that this employment was not assigned to individuals. Had such been the case, Gilbert would not have had the chagrin of seeing himself abandoned in the course of his laborious mission; agriculture would not have to regret one of its most enlightened supporters; and France whould have possessed five-thousand five-hundred Merino sheep, with the increase arising from them to the present time.

Gilbert had bought a thousand sheep in Spain, which entered France since his death. Part of this flock was left at the Rambouillet establishment; and the rest were distributed among those, who signed the subscription opened by government.

A considerable time had since elapsed, and no measure having been adopted for obtaining the 4500 sheep, which were so easily to be acquired, I thought that such favourable circumstances as then existed ought not to be missed, lest difficulties should occur, which might for ever deprive France of so valuable an acquisition; more especially after the term prescribed by the treaty of Bâle was expired. It was in consequence of these reflections that I thought it my duty to lay before my colleagues in the agricultural society of the department of the Seine, a proposition, which I thought calculated to meet the end in view. This being approved, the society addressed Government, praying for a transfer of the right to import these sheep, in order that a committee might be appointed to direct the proper measures.

The petition was acceded to, and the society immediately invited its members and other breeders to subscribe for the purpose of forming a company, which would take upon itself the charge of purchasing and conveying this flock to France. The company was organized; and sent out to Spain two agents, with several shepherds, to secure the valuable acquisition; but the Spanish government would not this year permit more than 1200 sheep to leave that kingdom. Part of these were sold in different departments, and the rest brought to Paris. The company holds itself pledged to introduce the rest.

Thus the number of animals carrying superfine wool, which are about to become the property of France, added to the million of pure or highly improved sheep already existing in different parts of it, will at least make it easy for the farmers to improve the wool of the native breeds, which is, in general, less in quantity, and worse in quality. This amelioration, which present circumstances favour, will, without doubt, make rapid progress.

Nothing can better prove the truth of my positions than the following simple calculation, suggested by my colleague Huzard, in his note to Daubenton's work, intitled ' Instructions to Shepherds:' " If the million of pure Merino and deeply crossed sheep, which France at present possesses, yield 5 lbs. of wool each in the yolk (which is a very low calculation), this same wool would weigh two millions of pounds, calculating the decrease by scouring at three fifths. These two millions of pounds of scoured wool will be manufactured into a million ells of cloth, with which five-hundred-thousand men may be clothed, at the rate of two ells per man. Is not this a considerable supply for our manufactures, grown in France, and independent of foreign supply ?"

Hence the most enlightened agriculturists of France, alive to their real interest, with which the prosperity of the country is closely combined, have proved at the last Rambouillet sales, the importance attached to fine-woolled sheep. Government has taken measures favourable to the progress of this pursuit by premiums, which will be as beneficial, in their results, to agriculture, as to commerce. Still it must be acknowledged with regret, that there exists in France, and particularly in the large towns, a certain indifference, which continually arrests our manufactures in their course towards that object, at which they aim.

Let us hope that government, that the men in power, and that rich land-owners, who have, in every country, so great an influence on the public spirit, will at length comprehend what it is their real interest to support by their example—the agriculture of France. Let us hope that those among us, who are still possessed of the *Anglo-mania*, will henceforth imitate the English in nothing but the love of their country, without which a nation cannot prosper; and that they will attach to the fabricks of France, the same price, which an Englishman affixes to the produce of his manufactures. Let us hope that, in future, we shall frequently read in our journals a paragraph similar to the following one, which is extracted from an English newspaper, dated 14th October, 1798:

" A ball will annually take place in the city of Lincoln, for the encouragement of the wool trade. Ladies will not be admitted, unless wearing stuff gowns and petticoats from wool, spun, woven, and manufactured in this country. Gentlemen will not be allowed to appear unless in woollen dresses, with the exception of stockings only : silk and cotton are proscribed."

Lord Somerville, formerly president of the board of agriculture established in England, has, in this respect, set an example worthy of being followed by our senators, counsellors of state, and all those who are paid by the nation: for surely these salaries should induce, as their consequences, the encouragement of French industry. " I am resolved," says his Lordship, " never again to wear superfine cloth, or kerseymere, any part of which shall be of foreign growth."

These patriotic sentiments, when combined with the assent of French gaiety, will form a spectacle, which is worthy of being exhibited by a great nation to the universe.

CHAPTER VIII.

HOLLAND.

THERE are few regions of Europe, possessing a temperature and soil, which differ more than those of Spain and Holland. The Merino sheep, transported from a scorching climate to a cold and marshy country, have, nevertheless, preserved, in Holland, the qualities which distinguish them from other breeds, and have remained vigorously healthy.

The first fine-woolled sheep, which came from Spain, met with the same fate, which attended them in every other part of Europe when first introduced; that is, they degenerated through the negligence of the breeders. Hence some animals, imported prior to the year 1789, have left a posterity no longer boasting the original excellence of fleece.

Attention to this breed is of no longer date than the above period, when Mr. Twent, who favoured me with most of the information which I am about to communicate, imported from Spain two rams and four ewes. These he placed on his estate at Raaphorlt, situated between Leyden and the Hague. The soil of this country is in part sandy, and consists of downs, on which the herbage is scanty. There are also in the district, meadows, and lands under culture, with others, principally growing forest trees and coppice wood.

The Spanish sheep, which had suffered much during the voyage, soon recovered from their fatigue, and became easily reconciled to their new food and climate. Their progeny, from that period to the present, forms the flock now on Mr. Twent's farm, which amounts to two hundred animals. He sells the surplus, because his farm is not capable of sustaining more than the above number. It is by parting with the least perfect animals, and preserving those, which bear the longest, as well as finest wool, that he has formed a valuable flock; preferable indeed to any in Holland. The rams yield 9 to 12 lbs. of wool in the yolk, and the breeding ewes 6 to 9 lbs.; a produce which far surpasses in value that of the best indigenous flocks, as the average fleece of these does not exceed 8 to 10 lbs.; and the wool of the Spanish race is sold for twice as much as the other. The Merino wool loses half its weight in scouring, and, since 1791, the cloths made it yield in no respect to the finest article, manufactured from the piles of Spain.

Mr. Twent, to establish the quality of the wool in his flock, made a trial, which I myself have adopted, in order that I might obtain positive proof of the respective qualities in the Rambouillet produce, as well as that of Sweden and Spain. The following is an extract from his letter:

" The wool of my Merino sheep has, hitherto, not degenerated, of which I beg leave to send the proof. I fixed on a piece of black cloth nine specimens of wool from nine young rams, to which I added one specimen of the best superfine wool from Spain, which a manufacturer could supply me with. After having numbered the samples, I noted them on paper, placing opposite to each number an indication of the animals, which had produced the wool. This memorandum I sealed, and sent to the merchant who had bought my clip, requesting that he would inform me to which wool he gave the preference. After having examined the specimens with attention, and formed his opinion, the paper was opened, when it appeared that five of my samples had been preferred to that of Spain."

Mr. Twent, who pursues the improvement of fine wool in Holland with much zeal and intelligence, has not been merely satisfied with establishing a pure Merino flock, but has tried the cross of this race on the different breeds of the country, such as those of the Texel and Friesland, which will be described in the second part of this work. The intermixture of these two breeds appears to promise great advantages. I have lately received letters from this agriculturist, in which he assures me, that the sheep of mixed race carry wool as fine as that of the Merinos, and even superior. He has observed, however, that the former, when pastured on rich lands, produce wool not so fine as when they grazed on a light soil.

Having been requested by the Economical Assembly of the Nation, at Harlem, to import another stock of Merinos, Mr. Twent obtained in 1792 three rams and four ewes through the Spanish ambassador.

During the same year, Mr. Cuperus procured some, of which I saw the descendants on his farm near Leyden, where they were kept during the fine season, on very rich and humid pastures. The race here obtained by the cross of the Spanish ram on native ewes, produced

H

wool, which almost equalled in fineness that of the pure Merino.

Mr. Kops also keeps, on the downs near Harlem, a flock of sheep bred from a Spanish ram, and the ewes of Friesland and the Texel. This new race, which had arrived at the third generation when I saw it, appears to promise happy results. The wool has produced cloths of very good quality.

The flocks of Mr. Twent and Mr. Cuperus have supplied rams to those individuals, in different parts of Holland, who have wished to improve the native breeds.

Mr. Collot has, at present, on a farm near the Hague, a good flock of the mixed breed; and the public spirit, which appears through Holland, as well as among all the civilized nations of Europe, in favour of this improvement, gives reason to believe that the Merinos will soon cause the common breeds to disappear.

CHAPTER IX.

CAPE OF GOOD HOPE.

I SHOULD commence this chapter by quoting a passage from Lord Somerville's Address to the Board of Agriculture, on the subject of Sheep and Wool, did I not think it better to leave his Lordship's detail unbroken. The reader will, therefore, find this account, among other matter of the greatest importance and interest, under the head of Great Britain.

At present, let the following particulars suffice. The indigenous sheep at the Cape of Good Hope are the breed with large tails, the one most generally spread over the globe. They produce wool which is particularly coarse, and resembles the bristles of the hog. The tail is very thin at its extremity, and turned up. The other parts of Africa, including those countries which border on the Mediterranean, produce only coarse wool, though some authors pretend to have ascertained the reverse in Morocco. It is probable that fine-woolled sheep existed in that country when the Moors cultivated the arts, and caused agriculture to flourish; but the breeds of sheep have been debased like the character of the inhabitants, under the domination of barbarous, ferocious, ignorant, and superstitious tyrants.

The best agriculturists, and those who are most employed in adopting and promoting the Merino breed, are of opinion that sheep, which are taken from north to south, that is to say, from a climate less warm than that to which they are transported, yield wool inferior to their former produce, and beget the same propensity whether allied among themselves or with the sheep of the country; and that those which pass from south to north, constantly improve the breeds with which they are united.

There are many existing facts which contradict this opinion; and it is easy to prove that the degeneracy which has taken place, is ascribable to other causes than those assigned. When a sufficient number of experiments shall have been comparatively made by accurate observers, it will be found that the want of intelligence and care, the bad choice of individuals for breeding, and the deficiency of proper food, have as much or more influence on the degeneration of any species: but this is not the place to discuss the matter.

I will content myself with remarking, that the success of the Merino breed at the Cape of Good Hope, proves the opinion so generally entertained, not to be always founded in fact. I am convinced, indeed, after the observations which I have collected in Spain, upon the breeds of that country, upon their mode of rearing the animals, upon the nature of the soil and climate, that the general causes of their fine wool are not those usually supposed. The preservation of the Merino race in its utmost purity at the Cape of Good Hope, in the marshes of Holland, and under the rigorous climate of Sweden, furnish an additional support of this my unalterable principle: *Fine-woolled sheep may be kept wherever industrious men and intelligent breeders exist.*

CHAPTER X.

ITALY.

Is Italy then, which has so long despised the useful arts, willing to awake from the trance, into which ignorance and fanatacism have till now plunged her? Has the contact with the French, hitherto so fatal, produced an electric movement, which leads her to objects of real utility? This wish a philanthropist might form; but wishes of that kind are seldom realized.

Piedmont, now an integral part of France, possesses many flocks of the pure Merino and improved breeds. I have received some specimens of wool produced by them, and I have found in them no deficiency as to beauty and fineness. I shall trace the history of the Merino naturalization in Piedmont, from a memoir, presented to the Academy of Agriculture.

Count Granerie, a man of genius and a sound patriot, a warm protector of arts and commerce, becoming a member of administration on his return from an embassy to Spain, conceived the project of securing to Piedmont this source of wealth, which now forms a principal occupation among most European governments. He obtained from the court of Madrid permission to take away a flock, which consisted of one hundred and fifty capital Segovian sheep, selected by the Prince of Masseran. Part of them were placed on the domain of La Mandria, and the rest were given to individuals.

The war, which prevailed at this period, did not allow the government to pursue the progress of this new establishment; and the loss of the minister would have been followed by that of the Merinos, had not the Academy of Agriculture, assisted by spirited and enlightened patriots, preserved this valuable depôt, by a sort of miracle, in its full purity of blood; since which the number deduced from it amounts to five thousand, one third being of the genuine race, and the rest a mixed breed of crosses on Roman, Neapolitan, and Paduan ewes.

In this state of things, almost all the proprietors of separate flocks, having united themselves in a pastoral society, proposed to government last year, that they would undertake the management of the La Mandria domain under certain stipulations, beginning with two thousand sheep which carried superfine wool; that they would increase this number by degrees to six thousand; and that they would regularly have a disposable number of rams, for the purpose of supplying the farms of individuals.

Exclusive of this establishment belonging to the society, mention must be made of the admirable flock belonging to Monsieur Collegno, which consists of one thousand five hundred animals. It is to his enlightened theory, confirmed by an exact and well-regulated practice, that the greater part of the success, attendant on this breed, is ascribable.—The other flocks of less consideration, when added to these, make a present total of five thousand, as above mentioned.

Let us now give an account of the two breeds. As to the pure Merino wool, it will be sufficient to observe, that neither in fineness, nor elasticity, nor strength, has it lost any thing; of which, indisputable proof is obtained by reference to the specimens taken from the sheep originally introduced. The form of the animals is exactly the same, and in height there is a gain of one to two inches. With respect to the quantity of wool, no comparison can be made, because when the flock arrived, it was not known how long a time had elapsed since they were shorn. Taking an average of the annual clip from rams, wethers, and ewes, the fleece may be calculated at 9 lbs., of twelve ounces. There are rams, which yield 16 lbs., and ewes, which reach twelve.

The flesh of the Merino sheep is infinitely

more delicate than that of the indigenous breeds.

The remarks on the mixed breed will terminate the observations, which it is proposed to make.

1. The wool of this race has been materially improved since the first year. It is necessary, when observing that one cross has effected so much, to state at the same time that the wool, grown in this part of Italy, is the best known after that of Spain.

2. The sheep of the mixed breed carry far more wool than their dams, the latter being only covered on the back, and the rest of the body left almost quite bare. Their fleece also is one third less in weight than that of the Spaniard, though the carcase is larger.

3. Their skin begins to assume a rosy hue, similar to that of the original race.

4. The new race exhibits a distinctive character, which is analogous to the sires, and departs in conformation from the dams. The head and neck are shorter, as are the legs; and the body is more compact.

5. The second cross has confirmed all that was promised by the first.

6. The third and fourth crosses leave so little

to desire either in quantity and quality of wool, or similarity of form, that none but a very accurate observer could perceive the difference between these, and the pure breed.

The following are the regulations, which have been established for the propagation and maintenance of the Merino breed in Piedmont.

1. There shall be appointed a Conservator General of all the flocks.

2. There shall be, in each department, Inspectors under the immediate dependance of the Conservator General.

3. These Inspectors shall proceed, in each department, to enumerate the above sheep.

4. Constant Registers are to be preserved, containing the names of those who are owners of flocks; the quality and the number of sheep belonging to each; and stating whether the owners are proprietors of the soil, or farmers renting it.

5. No owner of sheep, be he of what class he may, can use rams of any other than the new breed.

6. The Inspectors shall rectify the conduct of all those who act in opposition to this; and

shall take care that any lambs from other breeds are cut before six months old.

7. In the month of March there shall be a day appointed for a general assemby of all the proprietors, shepherds, &c. at La Mandria de Chivas: Those, who cannot attend, must send persons vested with full power to represent them. The assembly shall have for its president the Conservator General, and shall make every necessary provision for the advantage, and increase of the new race. On this occasion every Shepherd shall be furnished with a declaration, on the part of his Inspector, giving an account of his conduct, and the state of the flock under his care; when those, who have in their favour the greatest number of votes, and shew the best specimens of improvement in their flocks, shall obtain distinctions, and rewards to be fixed upon by the Conservator General in the name of the nation. At this assembly, rams shall be distributed according to the applications made, and the reports of the Inspectors. The distribution shall be settled by drawing tickets; but the shepherds, who have distinguished themselves, shall have a preference; and arrangements are also to be made

in their favour respecting price, and mode of payment.

8. The Shepherds shall be furnished with a map of the route to and from their respective farms.

9. Rams shall be sent into different departments for the use of individuals, conformably to the reports of the Inspectors, founded on the number of ewes existing in their respective districts.

10. The sale of any ram, derived from the flock of La Mandria, is absolutely prohibited till he is incapable of further service. The person, possessing him, shall not part with him till he has established the fact to the conviction of the Inspector, and received a permit. In case of death, the head of the animal shall be sent to the Inspector, to indemnify the possessor.

11. The compensation, which cannot be conveniently deferred till the period of general assembly, shall be adjusted by application, on the part of the Inspectors, to the Directors of La Mandria.

12. A register, containing the names of the shepherds and owners, to whom rams have been distributed, shall be kept by the Inspectors;

a duplicate of which shall be lodged at the domain of La Mandria.

13. The Shepherds are bound, on arriving in any department, to place themselves under the cognizance of the Inspector; and they cannot pass into another without a warrant, stating the number of rams, with which they have been provided.

14. The Inspectors will each month examine the state of every flock in their department, watching over the due execution of the regulations established, and making proper reports to the Conservator General.

Let me now be permitted to make a few remarks on these regulations, as far as they operate upon the preservation and increase of fine-woolled sheep in Piedmont.

It is evident that they were dictated by a sentiment of patriotism, for the purpose of accelerating rural prosperity and the manufactures of the country; yet they appear to me founded upon false and dangerous principles.

It is proposed to create a Conservator General and Inspectors of districts to superintend and execute the organization pointed out; but let it be remembered that the tribunal of the Mesta in Spain, and of the Foggia in Naples, which had an origin, similar to that of the Piedmont establishment, are become, by the lapse of time, the bane of agriculture and industry, in those two countries. Abuses, so pernicious as those sanctioned by the above codes, will, of course, never take place at a period, when the right, attached to property, is better known, and more respected than during the anterior ages; still, however, an institution of this kind affords, at every time, the means of usurpation and injustice. It would, therefore, be wiser to make laws, which extended no further than to secure the conveyance of the sheep, leaving the owners of flocks, or of mountain-pasturage, at liberty to make their bargains among each other, which would end in an adjustment of reciprocal interest.

The regulations, which compel owners of flocks to improve them by using Spanish rams only, are vexatious. They trench on the rights of property; and seldom attain the end proposed, because they can be easily evaded. Private interest will always operate more powerfully than public constraint. It is quite sufficient to make this interest comprehended, to issue good instructions respecting the management of sheep, to encourage the breeders, and to place at their door the means of improvement, which they themselves would be unable to procure.

The emulation, which these regulations tend to excite among the shepherds and flock-masters, is praiseworthy, and likely to be attended with good effects.

The modes of distribution or sale are bad, because they court intrigue, and are sure to be supported by favour. The auctions at Rambouillet are not attended with any of these inconveniences; and experience has proved that this method should be preferred to any other.

The measures, however, adopted in Piedmont, if we take them all together, will doubtless produce the happy effects, expected from them; especially if the project be put in execution, by which a manufacture of cloths is to be established in the district of Sacco; and proposals to this effect have already been submitted, by the Pastoral Society of La Mandria, to the Administrator General.

CHAPTER XI.

GREAT BRITAIN.

ENGLAND, which has, in these latter times, been distinguished by great improvements in agriculture, neglected, till within a very short time, the growth of superfine wool; partly because it was found so easy to obtain the article from Spain for the use of the manufactures; and partly because circumstances had not directed the attention of breeders to this species of amelioration. The long wools of good quality, not so common in Europe, and not less useful for certain fabricks, have had a preference in England; and industry has there produced a breed of sheep, carrying the most esteemed of all combing wool.

Prejudice has, in England, as well as in the rest of Europe, opposed the improvement of fine-woolled sheep. The breeders have fancied that the quality of the fleece was dependent on climate, soil, and pasturage; whence arose the erroneous supposition that the Merinos could not thrive in that Island; or at least that they would only supply wool of an inferior kind. The merchants and manufacturers, imbued with the same prejudices as the breeders and farmers, embraced a similar opinion; but they were led to it by different motives. They were afraid lest success should cause a diminution in the profits afforded them by exterior commerce.

The English, who have not known this breed in its living state much more than a dozen years*, have since that time imported a few of these valuable animals. When the interested and patriotic spirit, which animates this nation, is considered, no doubt can exist but that the Merino race will rapidly be spread through that country; and become a new source of wealth to a people, who so eagerly avail themselves of every source opened to industry.

The works which have been published by the Board of Agriculture, as well as the efforts of various societies and individuals, prove that the public spirit, in England, tends towards the encouragement of an object, so essentially connected with the increase, and prosperity of its manufactures.

The King of Great Britain, the Duke of

* This was written in 1802.

Bedford, a powerful patron of agriculture, Lord Somerville, president of the Board, and some other breeders, have procured Merinos, from which the stock is beginning to increase. It is gratifying to see the head of a government, as well as the men most distinguished by their influence, their wealth, and their knowledge, encourage, by all the means in their power, the most useful of the arts. It is gratifying to see a nation erect a statue, and strike medals to immortalize the memory of one, who has particularly distinguished himself in the walks of agriculture; as in the case of the late Duke of Bedford.

It will seem by the extract, which I am about to make from a work by Lord Somerville, what importance is now attached, in England, to the introduction of sheep carrying wool of superfine quality.

" All breeds of sheep in this kingdom may be arranged into two classes; those, which shear the short, or clothing, and those, which shear the long, or combing wool. The quality of the flesh in each class follows the character of the wool: the short-woolled sheep being close in the grain as to flesh, consequently heavy in the scale, and high flavoured as to the tase; the polled, long-woolled sheep 'more open and loose in the grain, larger in size, ' manufacturer's mutton,' fit for such markets, as supply collieries and shipping, but by no means, generally speaking, of such ready sale.

" Every practical man, looking over the map of England, who has given himself time to study the properties of its soil and climate, will admit, that one half the kingdom, at least, is by nature appropriated to the short-woolled, fine-grained breed. He might with safety admit much more than half. For it at length appears, that our climate, from the most northern parts to the most southern, can grow wool of the finest possible quality. Taking into consideration the upland pastures, the light convertible tillage, the loamy soils, and mountainous districts of the kingdom, such a proportion must be admitted to be moderate and just. But notwithstanding the great importance of the short-woolled sheep to the nation, as well in a commercial point of view, both as to the carcase and fleece, as with respect to the great extent of the kingdom appropriate to these breeds, the whole attention, both of farmers and breeders, has for these thirty years past been absorbed in carrying to a degree of perfection hardly credible, the heavy, long-woolled sheep; such as Lincoln, Cotswould, Romney Marsh, and new Leicester, but more particularly the last.

" To such extreme perfection has the frame of this animal been carried, that one is lost in admiration at the skill and good fortune of those, who worked out such an alteration. It should seem, as if they had chalked out, on a wall, a form, perfect in itself, and then had given it existence. Nay, fresh technical terms have arisen to express points in these sheep, thirty years ago unknown: such as the ' fore-flank,' and the ' cushion,' terms now universally admitted.

" Such is the animal now—almost the reverse of what it was. And from whatever source it originated, whether from the care and nice observations of breeders, or from crosses with Ryeland or Dorset flocks, is immaterial.—In eulogium of such, the author of this Treatise would have been gratified as to his own feelings; his doctrines too might have been received for a time with more popularity; but his duty to the public just then forbad it, and compelled him to take the part he did, hazarding doctrines unpopular with many superficial observers, with not a few, whose self-interested views it might derange, and with those, who hád the means to reflect with advantage, but who had not profoundly reasoned on a subject of such extreme national importance. All aimed at in this address, was, without partiality or indulgence, to impress on the recollections of farmers, that no breeds of sheep should be carried into districts ill adapted, both as to soil and climate, to receive them : that, in exertion to improve the carcase, they should not forget there was such an article as wool : that, in opposition to modern doctrines, the improvement of the one was not incompatible with the improvement of the other; and that the breed of sheep, which, on any given quantity of land, carried for a continuance the most wool as well as flesh, and both of the highest quality, was that breed to be preferred, of whatever description it might be, or from whatever country it might come. This was all the author ventured to suggest, and he would, under existing circumstances, have compromised the situation he then held, had he not done so. The delusion was too prevalent. It was a pleasant dream, and some did not like to be awakened.—Thus much in explanation as to what concerned the farmer. Let us now look to the manufacturer.

" Many of the fine cloth manufacturers, fancying, but without a shadow of reason, that it would be detrimental to themselves, wholly forgetting that they formed a part of that community whose interests they were bound to support, have laboured with no common pains to poison the minds of people in general on this subject; such we mean as from their pursuits could not either be well versed in trade or in husbandry; and for a short time succeeded: but who, by encouraging the wear of British cloths, would have given, in the outset, some little support to a national undertaking like this. Such manœuvres were unworthy British manufacturers, however for a short time they might succeed. It is not impossible, that, to do this the more effectually, some cloths have been sent to the London market, purposely ill manufactured. We would rather suppose they could not be so mischievously blind to their own interest; but such an idea must suggest itself, when we see the native cloths produced, worse in quality than those made long ago; such as hunter's cloths, and other sorts, known in the London markets. We have even been at a loss to conjecture from what cause our clothiers should set their faces against that improvement, by which every part of the nation must unquestionably derive such material benefit. There is not one well-grounded reason for the opposition shown to our endeavours. Were they all republican Frenchmen, they could have done no more. With pain we must reflect on it, but we refrain from indulging in that bitter invective, which such narrow policy has of late provoked, and content ourselves with remarking, that these gentlemen have fattened on the indulgence of government, and as is natural, indeed correct on such occasions, have been the first to fly in the face of its liberal and salutary measures.

" We speak not of all clothiers ; for many have paid every attention to this most important subject, and have conducted themselves with becoming liberality. But for those gentlemen before hinted at, who, it should seem, are little in the habit of reasoning deeply, we have only to suggest, that if they throw impediments in the way of such improvements in our wool, as their cloths in their manufactured state will best demonstrate (for ten facts are worth ten thousand fine speeches), we shall be compelled to employ those, who have all their masters' knowledge of the subject, and probably a great deal more, and want only the aid of capital to establish themselves in business with success.

" Advantage in the first instance was taken of the want of more explicit information by what means any material improvement was speedily to be obtained, and also of the author's subsequent absence from England. The operation on the fleece, by that mode of treatment recommended, although certain, must necessarily produce only a progressive effect. Had he made known his intention of crossing the seas for the purpose of bringing home a flock of Spanish sheep, his attempt must inevitably have failed, and might have risked the lives of those concerned with him. This attempt is not easily accomplished at any time, but is more than commonly difficult just now in a time of war, as well as from other causes needless here to dwell upon. It was an object not only to obtain the sheep themselves, but the whole system of management adopted by those, who had the care of these flocks of Spain. In both these particulars the author has been fortunate enough to succeed. The sheep are selected

from a *trashumante*, or travelling, *merino* flock
of undoubted high blood. The rams, twelve
in number, were picked from a flock of two-
hundred; for, except the *manso*, or bell-wether,
the males are left entire, on a presumption that
they carry more wool than wethers, and equally
fine. The couples (ewes and lambs) were selected from a number proportionably large. Of
the high blood and quality of this flock, the
admiration of those shepherds, through whose
flocks, twenty-two in number, they passed in
the course of their journey about the end of
March, was sufficiently indicative, if not otherwise well established. Of their properties as to
carcase, and, which is equally material, their
power of living hard, so as to carry much wool
and flesh on a small surface of land, it will be
as well to draw a conclusion from fact rather
than from assertion. That must not be deemed a bad breed of sheep, which, after a trial of
some centuries, can maintain its high quality
of wool, making two immense annual journeys,
and in a less space of time than could well be
supposed: more especially when we recollect
that the first journey commences, when lambs
are four months old, or even less. There are
few breeds in this country, that could support
such periodical marches for any length of time,

and not suffer materially in the form, as well as quality, of the carcase; for both are inseparable, being, as before observed, fed from the same sources." *

Although the success of the Spanish race in England be sufficiently proved by the above treatise of Lord Somerville, I think it, nevertheless, right to add the testimony, which is contained in a report of Sir Joseph Banks, relative to the flock of his Britannic Majesty. This paper is intitled, " A Project for extending the breed of fine-woolled Spanish sheep, now in possession of his Majesty, into all parts of Great Britain, where the growth of fine clothing wools is found to be profitable."

" After experiments had been tried for several years, by the King's command, with Spanish sheep of the true Merino breed, imported from various parts of Spain, all of which concurred in proving that the valuable wool of those animals did not degenerate in any degree in this climate, and that the cross of a Merino ram uniformly increased the quantity, and meliorated the quality of the wool of every kind of short-woolled sheep on which it was tried, and more particularly so in the case of the South-

* See Lord Somerville's System of the Board of Agriculture, p. 57, &c. 4to. 1800.

down, Hereford, and Devonshire breeds,—his Majesty was pleased to command that some Merino sheep should be procured from a flock, the character of which, for a fine pile of wool, was well established.

" Application was accordingly made to Lord Auckland, who had lately returned from an embassy to Spain; and, in consequence of his Lordship's letters, the Marchioness del Campo di Alange was induced to present to his Majesty five rams and thirty-five ewes from her own flock, known by the name of Negretti, the reputation of which, for purity of blood and fineness of wool, is as high as any in Spain.

" For this present his Majesty was pleased to give the Marchioness, in return, eight fine English coach-horses.

" These sheep, which were imported in the year 1792, have formed the basis of a flock now kept in the park of his Royal Highness the Duke of York, at Oatlands, the breed of which has been preserved with the utmost care and attention.

" The wool of this flock, as well as that of the sheep procured before from Spain, was acknowledged by the manufacturers who saw it, to be, to all appearance, of the very first quality; yet none of them chose to offer a price for it at all equal to what they themselves gave for good Spanish wool, lest, as they said, it should not prove in manufacture so valuable as its appearance promised. It became necessary, therefore, that it should be manufactured at the King's expense, in order that absolute proof might be given of its actual fitness for the fabric of superfine broad cloth; and this was done year after year in various manners, the cloth always proving excellent: yet the persons to whom the wool was offered for sale, still continued to undervalue it, being prepossessed with an opinion, that though it might not at first degenerate, it certainly sooner or later would alter its quality much for the worse.

" In 1796 it was resolved to sell the wool at the price that should be offered for it, in order that the manufacturers themselves might make trial of its quality, although a price equal to the real value should not be obtained: accordingly the clip of that year was sold for 2s. per pound, and the clip of the year 1797 for 2s. 6d.

" The value of the wool being now in some degree known, the clip of 1798 was washed in the Spanish manner, and it sold as follows:

" The number of fleeces of ewes and wethers was 89 ;

Which produced in wool, washed on

the sheep's backs, - - - - - 295 lbs.

Loss in scowering, - - - - - - 92

Amount of scowered wool, - - - 203

Which produced, Raffinos, 167 lbs. at 5s. per lb.

              Finos, - 23   at 3s. 6d.

              Terceros, 13   at 2s. 6d.

                                      ——————

                                   47l. 8s. 0d.

" The clip of 1799 was managed in the same manner, and produced as follows :

" The number of fleeces of ewes and wethers was 101 ;

Which produced in wool washed on

the sheep's backs, - - - - - 346 lbs.

Loss in scowering, - - - - - - 92

Amount of scowered wool, - - - 254

Which produced, Raffinos, 207 lbs. at 5s. 6d. per

              Finos, - 28   at 3s. 6d. lb.

              Terceros, 19   at 2s.

                                        ——————

                                   63l. 14s. 6d.

" The rams' wool of the two years sorted, together produced as follows :

Quantity of wool washed on the sheep's
   backs, - - - - - - - - - 314 lbs.
Loss in scowering, - - - - - - 99
Amount of scowered wool, - - - 215
Which produced, Raffinos, 181, at 4s. 6d. per lb.
               Finos, - 22, at 3s. 6d.
               Terceros, 12, at 2s. 0d.

                            45l. 15s. 6d.

" It is necessary to account for these extraordinary prices by stating, that in the year
1799, when both sales were effected, Spanish
wool was dearer than ever it before was known
to be; but it is also proper to add, that 5s. 6d.
was then the price of the best Spanish piles;
and that none were sold higher, except, as it
is said, a very small quantity for 5s. 9d.

" The King has been pleased to give away
to different persons, who undertook to try experiments by crossing other breeds of sheep
with the Spanish, more than one-hundred rams
and some ewes. In order, however, to make
the benefit of this valuable improvement in the
staple commodity of Great Britain accessible to
all persons who may chuse to take the advantage of it, his Majesty is this year pleased to
permit some rams and ewes to be sold, and also

command that reasonable prices shall be put
upon them, according to the comparative value
of each individual; in obedience to which it
has been suggested that five guineas may be
considered as the medium price of a ram, and
two guineas that of a ewe; a sum which it is
believed the purchaser will, in all cases, be able
to receive back with large profit, by the improvement his flock will derive from the valuable
addition it will obtain.

" Though the mutton of the Spanish sheep was
always excellent, their carcases were extremely
different in shape from that mould which the
fashion of the present day teaches us to prefer;
great improvement has, however, been already
made in this article, by a careful and attentive
selection of such rams and ewes as appeared
most likely to produce a comely progeny; and
no doubt can be entertained that, in due time,
with judicious management, carcases covered
with superfine Spanish wool may be brought
into any shape, whatever it may be, to which
the interest of the butcher, or the caprice of
the breeder, may chuse to affix a particular
value.

" Sir Joseph Banks, who has the honour of
being entrusted with the management of this

business, will answer all letters on the subject of it, addressed to him in Soho-square. The rams will be delivered at Windsor ; the ewes at Weybridge in Surrey, near Oatlands.

" As those who have the care of his Majesty's Spanish flock may naturally be supposed partial to the project of introducing fine wool in these kingdoms, it has been thought proper to annex the following notice, in order to show the opinion held of a similar undertaking in a neighbouring country, where individuals, however they may have mistaken their political interest, are rather remarkable for pursuing and thoroughly weighing their own personal advantage in all their private undertakings, and for sagacity in seizing all opportunities of improving, by public establishments, the resources of their nation.

" FRENCH ADVERTISEMENT.

" On the 24th May last, an advertisement appeared in the Moniteur, giving notice of a sale of two-hundred and twenty ewes and rams of the finest woolled Spanish breed, part of the flock kept on the national farm of Rambouillet: also two-thousand pounds of superfine wool being the present year's clip of this national flock; and one-thousand three-hundred pounds of wool, the produce of the mixed breeds of sheep kept at the menagerie at Versailles."

This advertisement, which is official, is accompanied by a notice from Lucien Bonaparte, minister of the Interior, as follows:

"The Spanish breed of sheep that produce the finest wool, introduced into France thirty years ago, has not manifested the smallest symptom of degeneration: samples of the wool of this valuable flock, which was brought from Spain in the year 1786, are still preserved, and bear testimony that it has not in the least declined from its original excellence, although the district where these sheep have been kept is not of the best quality for sheep-farming; the draughts from this flock, that have been annually sold by auction, have always exceeded in value the expectation of the purchasers, in every country to which they have been carried, that is not too damp for sheep.

"The weight of their fleeces is from six * to twelve pounds each, and those of the rams are sometimes heavier.

"Sheep of the ordinary coarse-woolled

---

* This must mean fleeces unwashed, or in the yolk, as it is technically termed.

K

breeds, when crossed by a Spanish ram, produce fleeces double in weight, and far more valuable, than those of their dams; and if this cross is carefully continued, by supplying rams of the pure Spanish blood, the wool of the third or fourth generation is scarcely distinguishable from the original Spanish wool.

" These mixed breeds are more easily maintained, and can be fattened at as small an expense as the ordinary breeds of the country.

" No speculation whatever offers advantages so certain, and so considerable, to those who embark in it, as that of the improvement of wool, by the introduction of rams and ewes of the true Spanish race, among the flocks of France, whether the sheep are purchased at Rambouillet, or elsewhere; in this business, however, it is of the greatest importance to secure the Spanish breed unmixed, and the utmost precaution on that head should be used, as the avarice of proprietors may tempt them to substitute the crossed breeds instead of the pure one, to the great disappointment of the purchaser.

" The amelioration of wool at Rambouillet, has made so great a progress, that in a circle from twenty-four to thirty-six miles in diameter, the manufacturers purchase thirty-five thousand pounds of wool, improved by two, three, or four crosses. Those who wish to accelerate the amelioration of their flocks by introducing into them ewes of this improved sort, may find abundance to be purchased in that neighbourhood at reasonable rates."

Such are the accounts, which I have obtained from a perusal of works on rural economy, published in England. I wrote to a gentleman in that country, for the purpose of obtaining more circumstantial details as to the present state of the Merino breed; and though I have not directly received answers to the questions which I sent, I have had the satisfaction to find those questions with replies to them on the part of Mr. Arthur Young, in the Bibliothêque Britannique. I shall here insert them.

Letter from Monsieur Etienne Vowet to
Mr. Arthur Young.

" A few days ago I received from my correspondents at Paris the following questions relative to Merino sheep. As I am totally unable to answer them, I address myself to you, sir, requesting that favour."

K 2

Questions.	Answers.
1. At what different periods have Spanish sheep been brought to England, and at whose expense?	1. It is about twenty years since the King imported Spanish sheep into England. Since that time others have done the same.
2. Have the descendants of these sheep preserved the fineness, of wool, by which the race is distinguished, or have they degenerated?	2. and 3. The fineness of the wool has been perfectly preserved.
3. In the latter case, to what should the degeneration be attributed? Is it to want of proper care and food?	
4. About what number of pure Merino sheep are there at present in England; and who are the owners of those flocks?	4. The number of Merino sheep in England is considerable. There are some in almost every district of Great Britain.
5. Have the native breeds been mingled	5. Various crosses have taken place, and

Questions.	Answers.

with the Spanish by crossing; and if so, what advantages have resulted?

with most success upon the Ryland, Mendip, and South-Down breeds. The improvement in the fleece has been great, but the sheep are deteriorated in carcase.

6. Have the improved breeds maintained their reputation?

6. Yes.

7. How many pounds of wool do the Spanish sheep produce, and what is its price?

7. The quantity varies from 3 to 5 lbs. It has been sold as high as five shillings per pound.

8. Is the food of the Spanish sheep the same as that of the indigenous breeds?

8. Yes.

9. Of what does it consist?

9. Sound pasturage and hay.

10. Are they left in the fields throughout the year?

10. Generally; but in some districts they have the shelter of sheep-houses.

134

Questions.	Answers.
11. What other care is bestowed upon the Merino race, and in what respect does it differ from that bestowed on the sheep of the country?	11. The care and treatment are exactly the same.

In asking the 7th question, I extended it to the mixed breed, but Mr. Young's reply is confined to the pure Merinos. I also asked whether the food, allotted to the Spaniards, was more expensive than that given to other sheep. The answer announcing it to be the same, we may conclude, as I have several times intimated, in the course of this work, that it is not more expensive to keep the Merino than any other breed, where flocks are allowed a proper quantity, and quality of support.

I have lately seen a synopsis of the fourteen breeds most esteemed in England for the profit which they return; which proves that the Merino race offers by its wool, much greater advantage to the breeder than any other; for in that country, where sheep occupy a great share of attention on every account, but principally on account of the fleece, an animal of the Lincolnshire breed, which yields the greatest profit in wool, will, when compared with a naturalized Merino at the present prices, pay only in the proportion of eleven to fourteen. *

* Since the period, at which our author wrote, this superiority of price, in favour of Merino wool, is become infinitely greater, and the present aspect of Continental affairs does not seem to sanction a probability of any material decline, for some time to come. I sold my clip of the present year (1809) at nine shillings per pound for pure Merino, six shillings for the different crosses of Merino-Ryland, and three shillings for unmixed Ryland, making an average of twenty-five shillings per fleece. The Ryland fleeces were, however, certainly not worth more than eight shillings each; and if reckoned only at that price, the average of all possessing Spanish blood would be considerably enhanced. The wool, when sold at the above prices, had been washed, according to the usual English method, upon the backs of the animals.　　　　　T.

END OF THE FIRST PART.

PART THE SECOND.

TREATMENT OF SHEEP.

CHAPTER I.

SWEDEN.

I BELIEVE that I have fully proved the success, attendant on the naturalization of Merino sheep in every country of Europe, which has made the attempt; among them, several materially differing from Spain as to soil and climate. It remains for me to shew, in this second part, the methods, pursued in various countries, with respect to the treatment and preservation of so valuable a race.

I have thought that, after combating the prejudices, and dissipating the doubts of breeders, it is not less important to exhibit the modes of treatment, which have had their advantages ascertained by the practice and experience of many good agriculturists.

In this exposition there will occur principles varying from each other in many respects, though the end of all is nearly the same; but this apparent dissimilarity, which is easily accounted for by the diversity of soil and climate, or even by the character of the people, will serve to establish two important truths, viz. that Nature arrives at the same end by different means, and that man obtains the same results with more or less certainty, in proportion as he more or less follows the track pointed out by her.

Those, who turn their attention to fine-woolled sheep, will adopt or modify the different methods, about to be mentioned, according to circumstances; but we believe that we shall offer some details little known, and the application of which will be useful towards the general health, and prosperity of their flocks.

I have before stated, that want of care and judgment has occasioned degeneration in many of the Merino flocks to be found in Sweden. I shall now, therefore, give an account of the system pursued by careful and intelligent breeders; a system, which cannot fail to secure the increase of this breed, in all its perfection.

The native sheep of Sweden are of a moderate size; the body long; the head small; the horns thin, short, and twisted backward; the tail shaggy, short, and destitute of wool at its extremity; the legs long and bare; the fleece open, of coarse quality, and of middle length. The ewes in general produce two lambs annually. In some parts, black flocks are to be seen; but the common colour is white. It is usual to shear these sheep twice, and even thrice, in the year. They have been generally mingled as to blood with the breeds of Germany, England, Eiderstel, and Iceland. The last-named race is the one, which best supports the rigour of a frozen climate, and exists on the most common aliment. These animals, which are never allowed any shelter either in Sweden or Iceland, where the snow often covers the earth for six or seven months, seek their food by scratching away the snow with their feet, cropping the branches of trees, &c. Their mutton is excellent. *

* See a former note.

Although these breeds are of a constitution more robust than that of Spain, though they can support themselves on food inferior in quality, and are less sensible of a rigorous climate, yet it has been, nevertheless, ascertained that the expense and care are much alike for both the breeds, if it be an object to keep both alike in good order. The flocks, which are neglected in the way just described, are in Sweden, as well as every where else, exposed to a great number of diseases; besides which they are liable to lose their wool, thus entailing a loss on the owner.

It is a custom of Sweden to keep the flocks in buildings which are too hot, and do not admit a free circulation of air. The enlightened agriculturists of that kingdom pursue a different method. They have spacious sheep-houses; and, even in the severest weather, cause their sheep to be driven out twice in the course of the day; having ascertained that fresh air is at all times salutary to them.

The sheep-houses have windows, which are kept open during the day; and the entrances are furnished, during fine weather, with hurdles instead of doors, that the air may have a free passage.

5

During winter, the sheep are entirely fed in these houses upon hay, leaves of trees, barley and oat-straw, peas-haulm and branches of hops. Each animal is allowed two pounds of hay, the rest of the meal being made up with straw or leaves. The two last substances are sometimes moistened with the residuum obtained at distilleries; and in some circumstances it is usual to allow a few oats, and chaff.

I shall here make known a mode of constructing racks, which I saw practised in several provinces of Sweden. This has the advantages of simplicity and cheapness to recommend it, while the inconveniences, resulting from the usual inclination of these fixtures, are avoided. Two common racks, or little ladders, are taken, and placed vertically on the ground, at the distance of a foot or eighteen inches from each other, where they are secured above and below. The bars must be near enough to each other, that the sheep may not put their heads into the racks; and the sides being vertical, the food will not fall on their necks and shoulders, as is usual when the common mode of construction is followed. In the latter case, the fleece is filled with dust and dirt, which clog the animal, and prevent perspiration. The particles of fodder adhere so closely to the fleece, that it is difficult to wash them out, or to be rid of them in any other process, to which the wool is applied. It is also well known that the sheep, when eating the hay or chaff, which is attached to them, at the same time tear off and swallow portions of wool, which are always unwholesome, and sometimes cause their death. Lastly, if the sheep be allowed to put their heads into the rack, they have a custom of stirring the fodder about with their noses, in order to find the most succulent part of it; whence it soon becomes infected by their breath, so that they refuse to eat it, and a loss, of course, takes place.

Folding is little practised in Sweden; partly because the farmers are not convinced that the system is advantageous; and partly from the fear of wolves and lynxes, which are too common in many parts of the country. From the latter circumstance it is usual to lodge all the sheep in their houses at night, even when the weather is perfectly fine.

During the day they are kept on fallows, stubbles, pastures, and occasionally meadows, both of natural, and artificial grasses. The natural herbage in Sweden is very short; but it is fine, thick, substantial, and of an excellent quality. The lands are inclosed by fences; and neither men nor dogs are generally employed for the preservation of the flocks.

•During hot weather, the sheep find shelter from the rays of the sun, under trees grown for that purpose on the farms; or they are conducted to the houses, where they enjoy a free salubrious air; it having been ascertained, in Sweden, that great heat is more pernicious to sheep than excessive cold.

The rural constructions in that country are raised some feet from the ground, and built on large blocks of wood; so that there is a space between the floor and the earth, in which implements of husbandry are deposited.

The sheep are watered every day at brooks, or other running streams; but they are never allowed to drink at a stagnant pool.

The best Swedish agriculturists consider salt to be very wholesome for sheep; which they give, more especially, in rainy and foggy weather. They often mix with the salt, wormwood or other bitter herbs, juniper-berries, and even pitch or tar. All these substances, reduced to a powder, and diluted in water, are administered in trunks of trees, hollowed for the purpose, and placed in the houses. Three or four small pieces of wood are attached to each of these trunks, standing vertically about a foot above it, to prevent any attempt at leaping on the part of the sheep, and to preserve the mixture from being spoiled by their dung, &c.

The composition, of which we have thus given an account, is regarded as a good preservative against many disorders in sheep, particularly dropsy, which is very prevalent in Sweden.

The owners of Merino flocks, who bestow care upon them, never use any sheep for generation till they are two or three years old, and cease to employ them after they have attained the age of seven. They do not suffer the ewes to be milked, as is the custom with the indigenous breeds; having remarked that the animals suffer from it, and that this custom not only injures the mother and lamb for the present time, but influences them afterwards, and alters their good qualities.

The shearing takes place towards the beginning of July. Some farmers only wash their sheep in running streams; others bestow more care on this operation. They place the animals in tubs, then wash them in a mixture of hot water and urine, finally cleansing them with pure water. After this process, the sheep are kept two or three days, or longer, if the season will permit, in a meadow, that the yolk may rise again, which imparts to the wool more softness and suppleness.

I will here just mention that there are few countries in which the art of forming fences is so various as in Sweden. While travelling through that kingdom, I saw not less than twenty kinds, differing more or less from each. other.

———————

CHAPTER II.

THE DANISH STATES.

THOUGH the Danes have not long turned their attention to the Merino breed, it may be useful to record the practice pursued with respect to them by the Messrs. Nelsons at Wiborg, where is to be found the flock originally derived from Sweden, and that since imported from Spain in 1797. The rural and veterinary acquirements of these two agriculturists have caused them to adopt a method, which experience has proved successful. I have seen but few flocks more carefully attended to, more healthy, and in higher order.

Sheep-houses have been erected for the Merinos, into which the light and air are freely admitted; both of these being, with great reason, deemed absolutely essential to the health of the animals. In fact, the sheep are guided by instinct towards the doors and windows, for the purpose of enjoying the light and air. The strongest are always found at this post, from which they drive their weaker companions.

The buildings are situated on some elevated spot, and face the south; they are furnished with windows, or apertures, in every direction. Within, they are divided by laths or planks, so as to have distinct stations for the rams, ewes, and lambs. Each division has, in its front to the south, a little separated yard, in which the sheep can walk, when the season does not allow of their being driven to the pastures.

There are double doors to each division, one of them being of lattice-work, which admits

L

air; the other solid, which is only shut when the weather is very severe.

The floor is well paved; and the boards, of which the roof is composed, are kept free from any interstice. The racks are made upon the proper vertical plan, which I have described in the preceding chapter; and at the foot of them are mangers, with small pieces of wood placed across them at short distances, in order that the sheep may not walk into them, or waste the food. These mangers are eight inches wide, and as many deep.

The sheep are fed in the houses on hay, or on rye and oat-straw, which is chopped before it is offered to them. This custom is generally practised throughout the North with respect to cattle in general, and merits attention, being economical, and attended with advantages. The Spaniards and Arabs give chopped straw to their horses; while it is erroneously supposed, in France, that mastication is better obtained by giving it in its natural length. Tenon, a member of the Institute, who has made some very interesting remarks on the teeth of the horse, proves that it is advantageous to chop the fodder for this animal.

The Merino sheep in Denmark are, at the

commencement of winter, almost uniformly · kept on straw and the leaves of trees. Towards New Year's Day, they are allowed a mixture of this with hay. Their food is administered three or four times in the course of twenty-four hours; they are made to move about in the yards; and care is taken that particles of their food do not fall upon their fleeces. Their allowance of fodder is about 3 lbs. per day each.

The native breed of Danish sheep bears a strong resemblance to that of Sweden. The size of the carcase is moderate; the head long and thin; the eyes small, the legs and tail without wool; the head and neck erect. The rams have small horns bent backwards, and the expression of their countenance is wild. In the western part of Holstein, and on lands which adjoin the sea, sheep are found, without horns, and having two pendulous skins under the throat. These animals are two feet eight inches high, and their wool, which is tolerably fine, is capable of being either carded or combed. They are only shorn once in the course of the year, and the fleece usually weighs about 8 lbs. The ewes produce annually from two to four lambs. The general weight of the race is 120 to 145 lbs.; and they are kept in the open air throughout the year.

Eiderstel, a country in the north of Holstein, on the western coast of the sea, boasts a breed, which, as well as the one just mentioned, has been employed to ameliorate those of Denmark, Sweden, and Germany. These animals are of tolerable carcase; they have no horns, and are distinguished by a tail, which is large at its junction with the body, and slender towards its extremity; they yield fine wool, but only a small quantity.

It has been observed, that where proper attention is paid to these breeds, and to the Merino race, the latter consume rather less food in proportion to their size.

The sheep are often left in the inclosures unattended by a shepherd; and when he takes them to the open fields, the flock is guided by a tame wether, which carries a bell on its neck. The method of the Spanish shepherds has been adopted, who never employ dogs to drive their flocks. Sometimes the sheep are fastened to each other in couples. This custom, generally practised in Denmark, is resorted to for the purpose of keeping them on the land, which they are at the time destined to graze.

While the rains are very heavy, the Merino sheep are confined in the houses; but great care is taken to keep the doors and windows open; at least when the weather is not particularly severe. Even in the latter case, only part of them are closed; a certain degree of cold not being pernicious to sheep, while the heat of their houses is always so. They are constantly permitted to walk in the yards, except when the snow falls in great quantities; they are never taken to low and marshy pastures; they are allowed a sufficient quantity of food, especially in winter,—an attention seldom bestowed on the native breeds; they are conducted into the shade during very hot weather; their houses are frequently cleansed, and proper measures are, in every respect, taken with regard to their preservation.

Salt is given to them, especially when the weather is damp, or when they are removed from green to dry food. Some persons give them the heads of herrings, or brine, in which meat or fish has been preserved.

The Danish sheep are shorn twice in the course of the year. This bad practice, which is attended with little profit, and exposes the animals to dangerous diseases, is not pursued with respect to the Spanish race. They are shorn but once; and during the operation, their legs are secured by list, or woolled garters, that they may not injure themselves by struggling.

The rams are not used till they are eighteen months old; and forty or fifty ewes are allotted to each. Experience has proved that rams can, without being enfeebled, serve this number of females annually, provided they are furnished during the season with wholesome and substantial nourishment. In like manner only eight to ten rams are allotted to a flock of four-hundred ewes at Neumunster in Holstein, where a good breed of sheep is kept. *

* Every foreign work relative to sheep, which I have read, seems to indicate that a much greater number of ewes is generally allotted to a ram in this country than in any other.—A neighbour of mine put one-hundred and twenty to a shearling, all of which were impregnated, and by good luck lived to produce one or more lambs each. Dr. Parry, to whose exertions, in judiciously mixing the Merino and Ryland breeds, England will for ever be indebted, employed, in the season of 1807, one of his most valuable rams to one-hundred and forty-six ewes. Lasteyrie himself afterwards mentions some instances of a still greater number; but he condemns the practice, and advises only thirty or at the utmost forty. Buffon goes no further than twenty-five to thirty. T.

A month before the rams begin their work, they are conveyed to the best pasturage, or kept at home on food of the best quality; but care is taken not to leave them together, lest they should, by frequent combats, lose a portion of their vigour.

The lambs are weaned at three months old, and taken to some good pasture at a distance from the ewes.

It is of great consequence to feed these animals well during their infancy, if we wish to fortify their constitutions against the train of disorders, which, in general, spread such devastation through our flocks. It is for want of abundant food, in the earlier stages of life, that sheep are often feeble, and of a degenerated appearance.

CHAPTER III.

SAXONY.

THE methods, pursued in Saxony, with regard to fine-woolled sheep, are certainly deserving of attention, whether we look to the Electoral farms, or to those of individuals; because the success of these methods has been demonstrated by an experience of nearly forty years, and the results leave nothing to desire.

This success is ascribable to the establishments originally founded by government at Stolpe, Hohenstein, and other places, where schools for shepherds were instituted; and to the instructions, disseminated through the Electorate as before mentioned.

Endeavours have been used, as far as soil and climate would permit, to treat Merino sheep according to the practice of Spain; they need no other attention than that which is founded on the principles, several times explained in the course of this work. We will now revert, however, to one point, in which a departure from the Spanish custom is adopted.

It is generally believed in Saxony, and other parts of Germany, as well as Holland, that the intermixture of fathers and mothers with their progeny, or even with that of animals derived from the same father and mother, will occasion degeneration in fine-woolled sheep. Under this conviction, the Saxon breeders often buy, from other Merino flocks, rams, which they substitute for those before employed by themselves; and land-proprietors bind their tenants, by a clause in the lease, to renew, every year, a certain number of rams. The undistinguished mixture, of the same flock, which has taken place in Sweden, France, and every part of Germany, proves in a decisive manner that it is useless to procure other rams, while any one possesses those, which are endowed with good qualities. If the opinion in question were correct, there would not be a single perfect sheep at this moment in Spain; for in that country the animals have copulated among each other for ages without distinction of parentage.

An owner of a flock will not only maintain it in a good state; but he will improve it if he selects sheep from another farm, which are more perfect than his own; but the improvement will arise solely from this increase of perfection of the animal acquired, not from avoiding intermixture of the same blood. The same end will be attained, though by slower progress, if the defective sheep of a flock are constantly discarded, and those only used, which possess, in an eminent degree, the qualities sought after. I repeat, that by choosing the most perfect animals, whether in his own flock or another, a breeder will be able to improve his stock; and he who makes the selection with judgment, supporting that judgment afterwards by the necessary care, will be sure to reach in time the end, at which he aims.

It appears singular enough, that long experience should not have convinced the Saxons how completely the degeneration of any breed was independent of consanguinity; but it was, of course, more easy and convenient to charge Nature with a defect, which had no other origin than in the neglect of those, who had the care of the flocks. I have, nevertheless, had occasion to converse with breeders, who were not of the erroneous opinion, which I here attack. *

The usual food, given to the sheep during winter, consists of hay, aftermath, trefoil, and oat or rye straw, &c. The hay is distributed twice or thrice in the course of the day, and in greater or smaller quantity, as it is more or less substantial. Those, who have no hay, substitute for it pease-haum, vetches, or lentils. Care is taken to cut the latter kinds of fodder,

* This subject has been lately discussed in a very intelligent letter from Sir John Sebright to Sir Joseph Banks.

T.

before their maturity, that they may be more nutritious, and that the fall of leaf may be prevented, which would, otherwise, take place for want of moisture.

Some farmers too make amends for want of hay by the use of cakes from oleaginous grain, by bran, and crushed corn, or sometimes meal. They mix the cakes and meal in vessels filled with water, which are placed in the sheephouses; and the residuum at the bottom of these vessels is afterwards given them to eat. I cannot too highly recommend this method, which contributes to preserve the animals in good health at a season, when it is difficult to procure fresh food. Grain given in this manner is more nutritious, particularly if the meal be mixed in hot water. This food is best adapted to the lambs; but when given to sheep, about six or seven pounds of meal are allotted to a hundred. When there is a want of provender, or the snow is of long continuance, corn is given to the sheep in the straw; or indeed sometimes, threshed and alone; but as the latter food is always expensive, it is generally soon replaced by roots of different kinds, viz. beets, turnips, carrots, and more especially potatoes. This method, hardly adopted at all in France, cannot be too strongly recommended to owners of flocks. It is well known that the dry food, on which sheep are obliged to live during a bad season, often occasions disorders; for which reason the English farmer cultivates turnips largely as his winter resource. Thus he is enabled to keep a larger stock than he otherwise could; a provision of roots being added to his ordinary fodder.

In Saxony great care is taken to collect the horse-chesnuts, which are regarded as a wholesome aliment and a specific against the rot. These are given to the sheep in autumn, when green food ceases. The chesnuts are cut into pieces, which it would be dangerous to omit, as they might, otherwise, stick in the throat of the animal, and cause its death. Sheep, as well as cattle, refuse at first to eat this food; but when accustomed to it, they seek it with avidity, and even like to eat the prickly husk, in which the nut is enveloped.

The wethers, and ewes, which have not had lambs, have generally no food but hay, or other inferior support; the best aliment being reserved for the ewe-mothers, the rams, and the lambs. The racks are constructed as before described; by which the fodder is not wasted, and the fleeces are preserved from injury.

During winter, the flocks are taken abroad when the season will permit. If the snow be not too deep, they are driven to the woods, or to dry lands and moors. Those breeders, who have no winter pasturage, keep their flocks in the sheep-houses from the beginning of November till April; but care is taken that they move about in the courts every day, and remain in the open air three or four hours. The doors of the houses, too, are frequently opened, that the air may be incessantly changed. There are even owners of flocks in Saxony, who have no pastures at all, and keep their sheep in the houses and yards throughout the year: nor is this treatment injurious either to their health, or the fineness of their wool, as long as care is taken to supply them with proper food, and to keep their houses airy.

The general custom, however, is to put them, during favourable weather, into pastures, where they find a sufficient quantity of food; and when these are not to be had, they are driven to the hills and other dry places. They leave the houses in the morning as soon as the dew entirely disappears, and repose in the shade, during the heat of the day.

When the rain falls heavily, or the fogs are thick, they are kept in their houses; nor are they suffered to go into the fields after an abundant shower of hail. In this respect the Saxons imitate the shepherds of Spain.

It is customary, on some farms of Saxony, to let the sheep drink in their stables during winter, instead of taking them to the watering-place. Spring or well-water is conducted by pipes into the troughs, and the animals resort to these when they like. Hence they drink often, and each time a less quantity, which is favourable to their health. The usual mode of watering sheep is attended with many inconveniences. The animals refuse to drink water in winter, when it is too cold; and during high winds, rain, hail, or snow, they hurry themselves, and do not take a sufficient quantity. They often disturb the stream too with their feet, which makes it disagreeable to them; and lastly, one part of the flock prevents the approach of another.

The Saxon breeders not only consider salt as a substance salutary to the sheep; * but are of

* I cannot omit this opportunity of recommending to the

opinion that it imparts a greater degree of fineness to the fleece. They administer it by mixing it either with the fodder, or drink. Some people mix it with hay-seeds, yarrow, bitter herbs, and a small portion of ashes. Salt is principally given during summer, when the weather is dry; and it is not allowed to ewes within four or five weeks of their yeaning-time; because it is thought such a provocative of thirst, that they drink too freely, and thereby make the birth painful; and that they feel so great a desire for it, as not to lick their lambs when they are dropped.

Such are, in general, the different methods of keeping flocks in Saxony; they are varied or modified according to the nature of the soil, and its products. The good farmers of that country observe the principle, without which no flock can prosper; that is, to keep a num-

perusal of every Merino breeder in Britain, the remarks on the use of salt for sheep, made by Lord Somerville in his excellent work, intitled, " Facts and Observations, &c." (3d Edit. p. 97.) and confirmed by his Lordship's most successful practice at an expense, which is small when compared with the advantages derived from it, high as is the price of the article. The convincing statement of his Lordship has induced me to follow his example.

T.

ber of animals only in proportion to the quantity of support, grown on their land. Experience has proved that the quantity of wool, produced by a flock, is always proportionate to the quality and extent of the nutriment which it has received.

The rams and ewes are not employed for generation till they have attained the age of two or three years. One ram is allowed to twenty or twenty-five ewes; and they are left together, day and night, during the season.

The period of admitting the ram to the ewes is determined, in different districts, according to the nature of the farms; the general object being to have the yeaning-season when Nature appears in renovated verdure.

The lambs, which fall before the month of March, are weaned in the course of June; and they are not afterwards permitted to join their dams in the pastures for some time. There are breeders, who keep them in the houses till autumn, and even throughout 'the first year; but care is taken to give them excellent food, viz. hay, aftermath, peas-haum, crushed peas and oats, &c.

The lambs are kept in the houses, because it has been remarked that those, which do not go into the pastures during the first year of their lives, are very rarely attacked by the dunt or giddiness. Breeders, however, are to be found, who allow the lambs to join their mothers a week after weaning.

When a ewe produces twins, she is only allowed to suckle one of them; the other is transferred to a goat; the milk of which animal has also been found wholesome for sick lambs; so that some owners of flocks keep goats expressly to be used in this way.

It is customary to cut the horns and the tails of the sheep. Shearing is not adopted on some farms, till the animals are two-toothed, it being thought that the lambs thrive better, if left in full fleece. Others pursue a contrary practice.

The shearing takes place at the beginning of May, after the fleece has been washed on the back of the animal. Formerly the wool was washed in warm water after being cut from the sheep, according to the Spanish plan; but this custom has been abandoned, in consequence of the wool felting into balls, by which its commercial value was reduced.

The mode of washing generally pursued, consists in driving the sheep through a brook

M

or rivulet; the next morning, they are again brought thither, and plunged into the stream, that every part of the fleece may be equally penetrated; after which the wool is pressed by the hand, beginning at the head, and proceeding regularly to the extremities. In the afternoon they are driven once more through the water; then two days are allowed for the fleece to become dry, and on the next, they are shorn.

A shearer dispatches twenty-five sheep per day. When an animal is wounded, the part is anointed with its excrement, or with a mixture of linseed-oil and rosin. The shearing ceases towards three o'clock in the afternoon, in order that the sheep may have time to feed on the pastures, to which they are taken after having undergone the operation. Some farmers fold their sheep for two or three weeks after shearing, where food is administered to them.

The wool is not sorted as in Spain, but it is thought sufficient merely to separate that part of the fleece, which is soiled by the animal's dung. The manufacturers find that the wool of Saxony thus washed, loses, in the scouring for manufacture, ten to fifteen per cent. more than that of Spain.

The shepherds in Saxony have no fixed wages; but receive a sum according to the profits of the flock. The owners find a great advantage in this arrangement; for the shepherds possess no means of deception, and are themselves interested in the welfare of the flock.

CHAPTER IV.

THE PRUSSIAN STATES.

THE sheep of the Prussian States yield, in general, wool of very coarse quality, which is shorn twice in the year, at spring and autumn. The Mark of Brandenburgh, nevertheless, contains a breed, which is only shorn once; and the wool is manufactured into good cloths, which are principally exported.

The wool of Silesia, as I shall hereafter mention, is of the best quality produced in Germany. Prussian Poland too possesses flocks yielding an article tolerably fine; but the sheep, on which Mr. Fink made his improvements, may be ranked, in their original state, among those, which supply an article of an inferior quality.

This agriculturist is not only the first in Prussia, who procured Merinos, and undertook the amelioration of the native breeds by the Spanish cross; but he is also the one, who, in all the country, has pursued, with the greatest care and perseverance, this part of rural economy. The method followed by Mr. Fink, and the principles, on which he acts, may be relied on with the greater confidence, because they are the result of forty-five years' experience.

I shall now communicate the information, which I have received from this gentleman himself, and from a little work, which he published not long ago in Germany.

He is of opinion, and with reason, that the fineness of the fleece is ascribable to the race of animals, which produce it, and not to any influence of climate or soil. Long experience has taught him that a breed may be kept, at the highest pitch of perfection, by chusing individuals, which possess the qualities desired; and it is by acting upon this principle that he

has imparted such perfect beauty to his flocks.
Having observed, for instance, that some animals carry, on a certain part of their body,
wool, which is finer than that grown on the
same part in other individuals, he has judiciously combined the crosses, so as to produce
the effect he had in view. This method, which
has been long pursued, with the intention of
producing perfection in the breeds of horses,
and which has supplied England with new kinds
of cattle, sheep, &c. should be carefully observed by all persons, whose aim it is to impart
a high degree of perfection to their wool-bearing quadrupeds. *

* Among the many and great improvements, which have
been effected by the judgment, skill, and perseverance of
Dr. Parry, may be mentioned the one alluded to in the text.
It has been a leading object with this breeder to procure,
as far as possible, a fleece of uniform excellence; and when
he did me the honour of sending a number of specimens
from his truly valuable flock, he was obliging enough to
take them from the shoulder and rump of each animal, to
exemplify his success. It would require much nicety to
discriminate between the produce of those two parts, often
differing so much in other sheep; and had they been shewn
to me without distinction, I am by no means sure whether I should not, in one or two instances, have given an
opinion in favour of the rump-wool.   T.

Mr. Fink, with a further view to the perfection of his flock, takes care to preserve thrice the number of lambs, which is necessary for his usual course of breeding. When they are two years old, he selects one third, and disposes of the rest. He is of opinion that food of a good quality, be its nature what it may, does not alter the fineness of the fleece; but merely causes an augmentation or diminution of weight, in proportion to its abundance and succulence, or the reverse. Bad or insufficient fodder is pernicious to the health of the animal, and consequently to the goodness of the wool.

The Merino breed and its crosses acquire a good size in Prussia, if constantly kept on succulent food; and this has been also ascertained at Rambouillet, and other places.

It has been remarked that air is favourable to the quality of wool; and that sheep, which are not kept long in the houses during winter, produce an article more proper for manufacture than those, which are confined throughout the inclement season.

Mr. Fink has also observed, that the degeneration of Spanish sheep (many examples of which occur in Germany) proceeds from a want of care in preventing the general admixture of these animals with the breeds of the country.

He has combated the opinion, generally received through Germany, that a race must degenerate when the sires copulate with their mothers, and brothers with their sisters. He thinks, nevertheless, that it is more certain to change the rams every third year, and even oftener, if animals can be procured, which carry finer wool, or even as fine as those in possession; but unless such is certainly the case, he conceives that those in the flock may be used not only for three, but four, five, or six years. He procured in 1779 three rams and twenty ewes bred in Spain, which he has kept free from any other connexion whatever, and which have produced sheep, constantly maintaining the original beauty of fleece. In other respects, he chuses rams for all his flocks, without any distinction, except that, which regards their individual perfections; having never observed that alliances *in-and-in* have produced the slightest degeneration.

The mode, in which he calculates the improvement of breeds crossed by the Merino ram, deserves to be here recorded.

The progress of amelioration is founded upon

Nature; and its effects are proved by the observations of Mr. Fink, as well as other agriculturists. It is true that Nature may swerve from her general track, and this sometimes actually occurs; but it is easy, by care and judicious selection, to bring her back; and again direct her efforts towards that point, at which every one aims, who endeavours to improve his breed.

Mr. Fink sets out by designating the ram which carries superfine wool, and is intended to improve any native breed, by the number 1
The coarse-woolled ewe is described by - 0

It is an established fact, that the ram has a much greater influence with respect to generation than the ewe; and consequently that the animals, proceeding from a fine-woolled ram, and a coarse-woolled ewe, will yield a fleece, which more resembles that of the father, than that of the mother; but we will, on this occasion, make our calculations from an unfavourable hypothesis, in order to prove our case more completely. We will suppose, then, that the qualities of the sire and dam are equally allotted to the offspring. Hence the lamb of the first generation will share the qualities of the father represented by 1, and the mother represented by 0, which makes it - $\frac{1}{2}$

The ram for the second generation is represented by - - - - 1

And the ewe, which has received one degree of amelioration as above, by $\frac{1}{2}$

The lamb derived from this connexion, will possess its father's properties in the proportion of - - - - $\frac{3}{4}$

The ram for the third generation is represented by - - - - 1

The ewe of the second generation put to this ram, by - - - - $\frac{3}{4}$

Whence the lamb, which is the offspring of this alliance, will be - $\frac{7}{8}$

The ram for the fourth generation will be - - - - - 1

The ewe of the third cross, and destined to produce the fourth, will be - $\frac{7}{8}$

The lamb derived from this connexion will possess the qualities of his sire in the proportion of - - - - $\frac{15}{16}$

The ram for the fifth generation will be, as at all times, - - - 1

The ewe of the fourth cross, to produce the fifth, will be - - - $\frac{15}{16}$

The lamb resulting from this will be $\frac{31}{32}$

and will, consequently, produce wool, almost equal in quality to that of the father.

It is certain too, that, in the course of subsequent generations, individuals will appear, carrying fleeces equal and even superior to the father; if choice be made of those, in every generation, which possess the desired qualities in the most eminent degree.

The six following rules are founded on the principles, which have been explained; and from these no departure can be made without amelioration being retarded.

1. He, who wishes to have a flock, bearing very fine wool, must, more especially at the outset of his enterprise, look for rams which have fleeces as fine as possible, in order that these may be used for the first generation; since it is evident that if the ram, employed for the second be better than his predecessor, time has been lost in effecting the proposed object.

2. The finer the original ewes are, the sooner will perfection be attained.

3. Attention should be paid that the rams, employed for future generation, are as fine as the first, without which improvement must be retarded.

4. If the breeder wish to attain a certain degree of fineness in the fleece, and proceed no further, it is easy. He may take rams and ewes of the first or second cross, which will possess one half, or two thirds of Spanish purity; and his flock will remain in this stationary degree of improvement for ever.

5. If the breeder be not attentive to the choice of his rams, and should he, for instance, put any common native ewe to a first-cross ram, instead of a pure one, the offspring will only possess one quarter of Merino blood.

6. If a ram, thus possessing only one quarter of Merino blood, be put to an unimproved native ewe, the lamb, resulting from this connexion, will only have one eighth of Spanish purity; and, by proceeding in this track, the breeds will be ultimately separated.

Mr. Fink keeps a flock of seven to eight hundred sheep, which are lodged in the houses during a considerable portion of the year, he having no winter-pastures. There they are fed on hay, aftermath, and chopped straw of different kinds. The species of food is changed as often as possible, because the sheep are found to eat, with superior appetite, in consequence of this variety; hence wheat, barley, and oat-straw, as well as peas-haum, succeed each other; but straw is never given without being chopped. Corn itself is only allowed when a very long winter, and consequent want of fodder compel the measure; except that a portion of it is always given to rams during the season.

Oat-straw is not often given to sheep in Prussia; it being there found too strong, and devoid of succulence. Peas-haum is preferred to any straw, when the stalk and leaves retain a little of their verdure; it is less esteemed when the stalk is become black, and the leaves fall off by threshing. These peas grow in the fields to the height of four, five, and six ells.

During winter the flocks are conducted to the fields of new-sown wheat and rye, while it freezes, and the ground is not covered with snow. They have straw given to them once in the day, merely to amuse them while in the houses, or to supply them with litter. This custom, which I have seen generally practised in Germany, is an excellent mode of economizing winter-food; and the kind of pasturage alluded to is also very wholesome for sheep.

The animals are removed to their houses at the beginning of November, and are not taken entirely abroad again till the end of March, when herbage begins to shoot forth.

The rams, which are usually kept with the young lambs, are now separated from the rest of the flock. Once in the day they are fed on the straw of farinaceous plants, and twice on hay; or, they have one allowance of straw, one of hay, and one of peas-haum. The old rams are also allowed oil-cake steeped in water.

The lambs have hay or aftermath twice in the day; and once peas-haum, or straw of some kind. They also are allowed the oil-cake dissolved in water, at the rate of 7 lbs. for one hundred animals.

The breeding-ewes are allowed a greater quantity of food than the other animals of the flock. They have daily one allowance of peas-haum, and two of other straw; or two allowances of the former, if the stock be sufficient to warrant this. When these aliments are not sufficiently nourishing or abundant, oil-cake is dissolved as above, at the rate of five, six, or seven pounds for a hundred sheep.

The wethers and barren ewes are usually kept apart from the rest. They are fed thrice in the day; twice with peas-haum, and once with straw; or twice with the latter, and once with the former, according to the stock of provender on hand; but they are never allowed any hay. Sometimes they receive a portion of the dissolved oil-cake above described; but this is only when the weather will not permit them to go into the fields of young wheat, and rye.

The breeding ewes are allowed, about three or four weeks before lambing, an allowance of hay, and two of different kinds of straw each day, with the oil-cake solution, at the rate of five or six pounds per hundred. At the period of yeaning, and while nurses, they have twice as much hay as straw; and the solution is improved by the addition of five, six, or seven pounds of meal. This mixture forms admirable nutriment, and causes the ewes to give a large quantity of milk. The meal is either from peas, beans, rye or barley.

The ewes begin to lamb towards the middle of March, and finish towards the middle of April, when they are conducted to the pastures; and brought back to the houses at noon and night, for the purpose of suckling their lambs. The latter eat aftermath in the houses, or, when the season permits, are occasionally taken abroad.

On the return of fine weather, Mr. Fink sends his lambs to the fallows till seed-time; when they are removed to pasture, or to the clover sown in the preceding year with barley or oats. Care is taken that the lambs do not eat too great a quantity of clover, before they are become habituated to this sort of food; and particularly that they are not admitted to it, while the plants are moistened by dew, or rain. This species of nutriment does not in the least disagree with them, when they are become accustomed to it, and eat it at a dry season. The clover-crops are sown thin, by which the plants are sooner dried after rain, and thus the growth of different herbage is facilitated, thereby forming a variety of food well adapted to the lambs.

These, as above has been observed, are only allowed the teat towards noon, and at night. At other times they are kept apart from the dams, it having been observed that they thrive better by this treatment. When with the mothers, they fatigue themselves by running; they are almost incessantly trying to suck; they refuse the herbage; and take less nutriment than when quietly kept on separate pastures. It is usual to mix a few ewes with the flock of lambs, for the purpose of directing them; or occasionally rams, when there are any of the latter, to which it is desirable that better food should be given.

Mr. Fink never shears his lambs, finding that he obtains as much wool by the omission, and that the young animals better endure the cold rains of autumn. Some agriculturists reject this method, asserting that the wool is thereby grown of unequal quality, and easily broken. In my opinion, the custom of leaving lambs unshorn should only be adopted in very cold and damp countries; or where proof exists that the animals are thereby preserved from giddiness, which some observations tend to shew.

Towards the end of autumn, herbage being not so abundant, and not so nutritious as before, Mr. Fink allows a supply of further food, sufficient to keep his stock in good condition. He has observed that frost, rain, and snow, impart a greater appetite to animals. He seldom folds them at the end of autumn; finding the trouble and expence of conveying litter for that purpose to be greater than if he kept the flock in the houses and courts appropriated to them, where they are always capable of being maintained in good condition.

He thinks that salt is not of much advantage to sheep, and need not be given but as a medicine. When it is perceived that, during a season of drought, they search with avidity for the salt-petre, which forms itself upon walls; he allots a small portion of salt to the flocks when they return to their houses at night; and on the following day he interdicts all means of drinking, and even moist pasturage. He conceives that if he suffered the animals to drink after eating salt, they would imbibe a quantity injurious to their health.

The method, pursued by Mr. Fink, with respect to his improved sheep, is the same which he followed as to the indigenous breed, before he entered on his system of amelioration. His only change is the solution of oil-cake in the water, drank by the animals. This fact proves that every breeder, who allows a common race of sheep the care requisite towards maintenance in a state of prosperity, will find but a very small additional expence requisite to keep a breed, which is gifted with extraordinary advantages.

The methods, pursued by some agriculturists

in Prussia, deserve to be known, as they are peculiar. Sheep are, in some districts, taken out every day, be the weather as it may. I have seen these animals in fields covered with snow, scratching with their feet to seek for food.

It is thought that heath is salutary, and that it does not deteriorate the fleece, when eaten with moderation.

In many districts of Prussia, the reeds, which grow in ponds, are given to sheep; principally the species called by Linnæus *Arundo angusti-folia.* The leaves of these reeds are accounted, when dried, a most wholesome sort of fodder, and often form the great resource for winter.

It has been observed that Merinos, after having been deficiently kept during winter, assume a ragged appearance and lose their wool in locks, as is the case with the native breeds.

The dogs, used in Prussia, for driving the flocks, are of a race differing from those known in France, under the name of shepherd's dogs. They are of a small compact form, and prick-eared, resembling our wolf-dogs. Some of them are close-haired, others are shaggy; but all are very docile, and never wound the sheep. At the voice of the shepherd, they repair to that part of the flock, which is pointed out;

and if the animals refuse to advance, the dogs push their noses against them, which is enough to make them adopt the desired direction. It is much to be wished that our shepherds would accustom their dogs never to bite the sheep. These animals, naturally timid, are quite alarmed when the dog approaches them, under our system; they press against each other, run away with terror, and are often materially injured. Their constant state of fear disturbs their repose, and makes it impossible for them to graze in tranquillity, which is prejudicial to their health. Few of our flocks are seen without individuals, at one time or another, exhibiting marks, which have been caused by the tooth of the dog.

Several breeders have assured me that they have secured their flocks from the dunt, or giddiness, by fastening a cloth, covered with pitch, upon the heads of the lambs before they were conducted to the fields; they assert that this species of cap prevents the introduction of the maggot to the frontal sinus which is lodged by the fly during the early period of the animal's life. It would be well to ascertain, in a positive manner, so important a fact.

I must not silently pass over another operation, practised by German agriculturists; the good effects of which have been thoroughly ascertained. I allude to inoculation for the disease known by the name of *claveau*, or sheep-pox; which produces the same effect upon sheep as inoculation for the small-pox on the human race. The animals contract the disease; but the symptoms are milder, and the recovery rapid as well as almost certain. Experience has proved that inoculated flocks are secure against contagion. Ewes, after that operation, have been put to rams infected by sheep-pox; and neither these ewes, nor their offspring, have shewn any symptom of the disease.

Inoculation is performed by making an incision within the thigh, upon that part which is without wool, about four inches from the anus. The incision should perforate the skin; but care should be taken that it does not injure the muscles, or occasion an effusion of blood. A pustule of an infected animal is then pressed by the fingers, and the virus, being taken from it on a lancet, is conveyed to the incision of the other.

Some experiments have been tried in France for the purpose of ascertaining whether vaccine inoculation would preserve sheep from the *cla-veau*. The results, hitherto obtained, appear satisfactory; and it is probable that this new preservative may be substituted for the one, which I have just mentioned.

SILESIA.

BEFORE I record the system of management pursued by the Count von Magnis, it is advisable to mention the native breeds of Silesia, and to give some idea of the mode, in which they are generally treated. The comparison of the two methods will be instructive, and will prove how far intelligence and experience operate upon the progress of rural economy. The astonishing improvements, effected by the enlightened nobleman just mentioned, will serve as answers to the sophistry which is incessantly brought forward by scepticism, and adopted by ignorance. They will prove that agriculture cannot prosper in any State, unless it be studied and practically pursued by the landowners; and they will evidently shew the necessity of disseminating knowledge among the general class of farmers. *

* This is evidently applicable to several districts of England, and, among the rest, to that, in which I live; where the farmers appear to despise the idea of any knowledge, which a book can convey, and laugh at all principles founded upon the basis of modern experiment; while they, with undeviating filial respect, follow the blundering track of their fore-fathers, and will, I dare say, pursue the same *jog-trot* system of ignorant folly to the end of their lives. I am most ready, and most happy to say, at the same time, that there are others of sentiments exactly opposite to these, anxious to obtain knowledge, eager to act upon it, and ever ready to share it with their neighbours; while nine of those neighbours, out of ten, are as ready to refuse the offer. Let Lord Somerville make a few brief remarks to these people. Let him first give the questions, which they constantly ask, and then return the answers, which so well deserve their attention. The questions, on the part of the farmers, are: " Why is it that these men should do more than we do? Why do not we draw from our management as much for the public supply?" The answers of their patriotic adviser are: " Because you are tied down, and fettered by old prejudices, adverse to your interest; and because they, profiting by a more extended education, reason on a greater scale, and act on principles, more consistent with nature and good sense."

Here too I may well quote some admirable observations by the Duke of Liancourt on the same subject. " How little progress" (says he) " would have been made in

The indigenous breeds of Silesia are, in general, of small carcase—the neck long, as are the legs, and the latter are devoid of wool. Such is also the case with the belly, and part of the head.

There are two varieties in this country, one of which is very inferior to the other with respect to the quality of the wool. The better

chemistry, in physic, in geometry, and, in short, in any of the sciences, if those, who devoted themselves to the study of them, had been contented to abide by their own discoveries, without deigning to gather further knowledge from the experience and researches of others? This species of communication is more necessary to farmers than to any other class of citizens ; for as the care and business of their farms entirely employ their time, they live within their own immediate circle, and are ignorant of every thing that passes out of it. The most minute and exact details are required to oblige them to consent to what they would call an innovation of the established rules of agriculture, and this can only be done by convincing them, as much as possible, of the utility of changing their usual plan ; and of adopting a newer method, which they would never seek, being satisfied with what use has habituated to them."

May the growers of fine wool near me, and every where, feeling the force of arguments like these, be soon convinced of the important facts, which imperiously demand (and more especially in the present state of affairs) their adoption of the Merino breed!   T.

race carries a fleece preferable to any in all Germany; and is found principally in the districts of Namslau, and Oels.

Sheep are shorn twice in the course of the year, though this is, in general, considered a bad practice; because it exposes the animals to dangerous accidents, while the wool is less adapted to the different operations of manufacture.

In those parts of Silesia, where sheep are most attended to, the hay is reserved for the breeding ewes, and the hogs. The other food consists of rye, wheat, or oat-straw, and peashaum. Mineral salt is given to the animals, and it is suspended by cords from the roofs of the sheep-houses, for the purpose of preventing the contusions, which would ensue, if general avidity were excited by putting it into the troughs.

Sheep are never folded in Silesia. They are housed at night even in summer; and remain in their habitations during a considerable part of winter. The treatment of them is incorrect; for the houses are not well-contrived, and the air is constantly impregnated with the vapours arising from the litter, which is only removed every half-year.

The buildings belonging to the Count von Magnis are, however, regulated in a very different manner, upon rational principles, and such as are conformable to the nature of sheep. The houses are built of wood, with apertures two feet square, and two feet from each other. These windows remain open almost throughout the year; and frequently in the severest frosts; care being only taken to protect the ewes from the weather at their lambing-season.

The breeding ewes, the wethers, the lambs, and the rams, are kept in separate buildings and courts.

These flocks, contrary to the general custom of Silesia, are only shorn once in the year; and the milk of the dams is entirely applied to the support of their lambs; for it has been observed that when the ewes are milked, the lambs are less vigorous, and their fleeces fail both as to quantity and quality. The winter, which sets in during October, and continues to the end of April, makes a much larger provision of fodder necessary than in other countries. Exclusive of this, the mountains, on which the estates of Count von Magnis are situated, consist of land, the soil of which is arid, and devoid of natural herbage. Hence, it has been necessary to employ every resource of art, for the purpose of creating artificial pasturage on a tract of country, which would hardly produce any indigenous plants.

By a culture, founded upon well-established principles, the Count has, nevertheless, succeeded in annually obtaining two-thousand five hundred quintals * of clover, and a proportionate quantity of lucerne, potatoes, &c. for the winter's food of his sheep, without reckoning the fodder necessary for a large herd of cattle, which he constantly soils in the stables, also in opposition to the custom of the country.

This great agriculturist has pursued a system, which ought to be imitated by all those, who aim at a perfection of culture; and which may be adopted, without much trouble, upon a soil of any nature—he has almost entirely banished fallows. By this step it is that he has procured

* While I was in Germany, I remember to have heard a quintal generally reckoned as one-hundred pounds; but Lord Somerville, who is much better authority than any that I can bring forward, states it as one-hundred and twenty-eight. His Lordship alludes, however, to a Spanish quintal, which may not be the same as in Germany.

T.

food for his numerous flocks, and immensely added to the mass of his revenue.

The lands, wich were formerly suffered to lie fallow, are now in almost one continued round of produce.

Part of these are sown with lucern and clover, which produce crops, not only sufficient for the support of the cattle during summer, but also of the sheep during winter. A third is devoted to potatoes for the winter-keep of the flocks.

The pastures are on land, not capable of growing corn, or at too great a distance to be used for that purpose. They are sown with white or Dutch clover (*trifolium repens*), saint-foin (*hedysarum onobrychis*), lucern (*medicago sativa*), oat-grass (*avena elatior*), burnet (*poterium sanguisorba*), and burnet-saxifrage (*pimpinella saxifraga*). These different plants are mingled in the same field, and form the pasturage of sheep during six or seven successive months. A new portion of this land is ploughed every year, and produces potatoes for the next crop.

That portion of the Count's estate, which used to lie in fallow before his new system was introduced, is now cropped as follows:

5

First year   -   -   Oats;

Second year   . -   Potatoes;

Third year -   -   Oats or Barley;

with which are sown the seeds above-mentioned.

Fourth year   -   Seeds;

which are twice mown.

Fifth year -   -   Seeds still;

which are now converted into sheep-pasture, and used as such for one or two years according to their goodness.

This mode of culture supplies an abundant provision of straw, potatoes, and other provender for the winter magazines.

The division of this fodder is made in a way, which secures to each animal sufficient nutrition, be the rigour and duration of winter what it may. After harvest, the Count makes an immediate calculation of the quantity of food, which he has obtained; and from these calculations he allots the quantity for each class of animals, and that, which is to be held in reserve.

In order that the distribution may be regularly made, as well as that it may be easy and exact, he fixes tables in each sheep-house, on which are inscribed the class of animals, the hour, at which they are to be fed, the nature and quality of the allowance which they are to receive, &c.

The following are two tables on his Lordship's established regulations:

For 125 Ram Lambs.

		Weighing lbs.
At six in the morning	{ 2 Portions of clover - - { 2 Do. of chopped straw -	} 80
At ten o'clock,	The same - - - -	80
At one o'clock,	Hay - - - - -	62
At four o'clock,	{ 3 Portions of chopped straw { ¼ of a portion of potatoes	} 48 } 120
At six o'clock,	Chaff - - - - -	92

making a daily allowance of 3¾ lbs. per head.

For 100 Ewes.

		Weighing lbs.
At six in the morning	{ 1¼ Portion of clover - - { 2¼ Do. of chopped straw	} 69
At ten o'clock,	The same - - - -	69
At one o'clock,	Clover - - - - -	100
At four o'clock,	{ 2½ Portions chopped straw { ½ Portion potatoes -	} 144
At six o'clock,	Chaff - - - - -	75

making a daily allowance of 4¼ lbs. per head.

The allowance of the ewe-hogs is fixed at 3¾ lbs. per head; and that of the rams, as well as the wethers, at 5¼ lbs. The Count von

Magnis gives an abundant quantity of food to his sheep, as may be seen by these calculations for each animal; and he endeavours to increase this allowance every year, by multiplying the artificial grasses on his estate. He very properly thinks that the quantity of wool depends entirely upon the quantity of nutriment.

He never gives corn to his sheep; for he has found that the place of this expensive aliment is well supplied by potatoes; and that they fully answer the same purpose at one quarter of the expense.

He allows his sheep to eat as much salt as they like; and the usual consumption in the year is about two quintals and a half for an hundred sheep.

It is by this excellent method that the Count constantly maintains his flocks in a state of prosperity, and that they repair to their pastures, after a long winter, in as healthy a state and as good condition, as when they quitted them to take up their abode in the houses.

He has procured foreign breeds at a great expence, and especially the best Merino rams, which were to be procured in Saxony, or the Austrian States. He takes particular care to number every sheep in his flocks, that no race may be mixed contrary to his intention, and that he may, with perfect exactness, establish the results of his experiments. This system is the more necessary with him, because he thinks, with the greater part of German agriculturists, that the breeds would degenerate, if care were not taken to avoid alliances of parentage. A number is branded on the neck of the lamb, when separated from its mother; or a distinguishing ear-mark is made. These numbers are inserted in a register, which also records the quantity of wool produced by the individual, the source from which it is derived, the experiment to which it is devoted, and the other observations, which concern it.

I do not exactly know the system of numbering, invented or adopted by the Count; but as such a mode of distinction may be useful to almost every breeder, and especially those, who adopt the Merino race, and who should, therefore, be attentive to the degree of blood in the rams and the ewes, which they employ; I shall, in the last chapter of this work, describe a mode of distinction, to be adopted by those, who like it; and which will, I conceive, be satisfactory.

In the method of washing wool, the Count

von Magnis follows the Silesian method. He causes the animals to be well washed in a stream, after having compelled them to leap into it from a bridge raised somewhat above it.

The different breeds, which he possesses, are only shorn once in the year. Each fleece is weighed immediately after the animal is clipped; and the weight, as well as the quality, is noted in a register, opposite to the number of the individual, which produced it. It is from his calculations upon these registers, that the Count makes his selections of rams and ewes, best calculated for each other. He combines different qualities, such as the fineness of wool, the length, the quantity, the size of carcase, symmetry, &c. and by these combinations he obtains animals more perfect, and more productive.

With all this care, attention, and perseverance, it may naturally be concluded, that the Count von Magnis has brought his flock to a very high pitch of perfection. Such would not, however, have been the case where the territorial impost is exorbitant, unjustly divided, and (more disastrous still) where it is not fixed, but varies every year according to arbitrary will; and constantly falls upon the farmer in proportion to the fruits, produced by the effects of his own industry. Land was taxed in Silesia about the year 1748, and no advance has taken place since that time. Silesia too is a country the most populous, as well as the best cultivated of any in Germany, and its industry is remarkable, though its physical state and position are unfavourable.

The example of Count von Magnis is, doubtless, worthy of admiration, and deserves to be followed. Men of limited fortune cannot, of course, reach improvement so prodigious and so lucrative; but there is no land-owner who may not, by the cultivation of his inheritance, proportionately increase his fortune, small as it may be, if he unites, with rural knowledge, the will, the intelligence, and the care, indispensable towards the success of his enterprise.

I have witnessed in England, and more recently in Germany, many modes of culture, brought to a high degree of perfection by enlightened individuals; but it is in vain to expect perfection on the part of simple and ignorant men who are incapable of attaching merit to any method of proceeding, which deviates from their accustomed routine.

Let those, who are in easy circumstances, abandon their towns and their palaces, where Ambition reigns, and Vice receives homage; let them, at length, deign to become instructors in the most useful of the arts; and France will soon see the public revenue increase, through the fortunes acquired by individuals.

CHAPTER V.

THE AUSTRIAN STATES.

It is an observation, which has been made, and a hundred times repeated by travellers, that the agriculture and industry of a country are more or less flourishing, and the people more or less happy, in proportion to the degree of liberty, which is enjoyed, or in proportion to the prejudices, by which the religion of the nation is degraded. This remark, which the mind is led to make when comparing the state of Italy and Spain with that of certain

other countries in Europe; when comparing
Switzerland with herself, and above all the North
of Germany with the central parts of it—this
remark, I say, appears to be proved more correct in the country last named than in any of
the rest.

In those parts of Germany, where the people have sound notions, ease and prosperity are
seen to reign; while in the states, subjected to
superstitious practices, the people, ignorant,
and more or less enslaved, are incapable of
concurring towards the progress of culture, and
other branches of human industry.

This truth is striking, and more particularly
in the Austrian States. Upper Austria may,
indeed, be excepted; for there the land is well
cultivated; but, in almost every country, peculiar causes unite to render agriculture and
industry prosperous in spite of the vices, which
disgrace its government. Thus the environs of
Naples, and of Valentia in Spain, are distinguished by good culture, and abundant crops.

But, when we speak of the Austrian States
in general, which include, besides that Duchy,
the kingdoms of Bohemia and Hungary, Moravia, &c. &c. agriculture may be said to lan-
guish; and the efforts which, at different pe-

riods, have been made to revive it, still continue fruitless as they have hitherto always been ; for agriculture, like commerce, cannot flourish unless it be sustained by liberty and directed by knowledge.

The vast territory of these States produces a great variety of native breeds, the wool of which is inferior in quality. The sheep are, in general, kept upon the worst pastures; they are not properly attended to; they are housed in confined buildings without apertures, from which the dung is only removed twice in the course of the year. This treatment, and bad food during winter, occasion frequent diseases in the flocks and herds. There are, in reality, few countries, where the murrain is so frequent and so disastrous as in the Austrian States. The custom of milking the ewes, and that of letting out a flock to some peasant, allowing him, in return, a number of the animals which compose it, are also measures, in opposition to the progress of improvement.

The milk of these animals is used to make butter and cheese. At Brunn in Moravia, a kind of cheese is made, which is held in very high estimation. It is called Brunsen, and a large consumption of it takes place at Vienna.

Hungary is that part of the Austrian States, in which the greatest number of sheep are kept. Here flocks are to be found of one to seven thousand animals. The indigenous breed of the kingdom is large in carcase, and produces wool, which is long, coarse, and of a waved appearance. Both the rams and ewes have horns, which are vertical and spiral.

The sheep of this race, which are also to be found in Bohemia and Wallachia, * are known to naturalists by the name of *ovis strepsiceros;* and appear to have been originally derived from the island of Crete. They are of robust constitution, and their skin, with its covering of wool, is employed to make dresses for the use of the peasantry. The shepherds anoint these

---

* In the translation of Buffon's Natural History, published by Barr (1792), two curious engravings of these Wallachian sheep are given, which strictly answer Mr. Lasteyrie's description. The Editor of the above version states, that he was favoured with the drawings by Mr. Colinson, a Fellow of the Royal Society in London ; that the representations were taken from two living animals, and that the likeness was strongly attended to. The horns, which vary in the two sheep, impart to both a most singular appearance. The prints are in the 5th Vol. of the above Edition, and are described at p. 263.   T.

skins with lard; and they are then used instead of a mantle during tempestuous weather, or when their owners sleep in the midst of snow.

Moldavia possesses a breed of sheep with large tails, which yield one of the longest wools on the Continent, if I may judge by the specimens received from that country. The fleeces of these animals are composed of two sorts of filaments; the one very coarse and about eleven inches long; the other of somewhat better quality, and only from three to seven inches long.

The wools, most esteemed in the Austrian States, are those of Moravia and Bohemia. The Moravian sheep, of which I have seen several, are very large. They are high-backed, have a small head, and carry long wool.

The amelioration of fine-woolled sheep has been more encouraged in Austria than other branches of rural economy; as will appear from the following regulations, issued by government.

1. The sheep, carrying superfine wool, are not to be grazed upon moist commons and plains.

2. For propagation, rams are to be used,

which are of the third cross, by Merino rams upon Bohemian ewes.

3. Rams of the Spanish and Paduan breeds are to be passed from one flock to another.

4. The circles of the Provinces will annually communicate to government, a statement of the flocks, and progress of amelioration, accompanied by specimens of wool.

Two years after the first regulation appeared the following ordinance, for the purpose of preventing the murrain (*epizootie*).

1. The superfine-woolled sheep are to have fodder of good quality, and in sufficient quantity.

2. They must not be taken to low-lands, or such as have been inundated.

3. The doors and windows of the sheephouses are to be opened before the animals are driven out; and when the weather is favourable, the windows are to be left open, day and night.

4. The flocks are to be taken abroad during fine weather, after having had food and water; and they are to be kept, for several hours, upon a spacious open tract of land, on which are no pernicious plants, or stagnant water.

5. In wet or cold weather, salt is to be given

twice in the course of the day ; particularly to those animals, which have reached their full degree of growth.

6. The litter must be regularly taken from the houses, and renewed every time that sheep leave them.

7. Care is to be taken that ordure, and other substances attached to the fleeces, be removed.

8. The animals are not to be taken into the pastures till they have had a sufficient quantity of food in the houses, with salt, and water.

9. They are to be kept at home till the herbage of spring has shot forth tolerably.

10. With respect to dead animals, and their skins, the advice is to be followed, which is given by Professor Wolstein in his work On the Murrain.

This work, of which there are a dozen editions in Germany, is intitled : A Treatise on the Murrain, which prevails among horned Cattle, Sheep, and Pigs. It is a most useful publication, and has been distributed through the chief principalities of Germany by order of government.

CHAPTER VI.

FRANCE.

IT was my wish, upon this occasion, to have detailed the system, pursued at Rambouillet, with respect to the treatment of Merino sheep; but the information which I applied for, on this point, was refused by the registrar of that establishment, who doubtless, in thus rejecting my application, obeyed the orders of those above him. I must, indeed, acknowledge the frank and honest manner in which he received me; but I must also express my astonishment at finding in a national establishment, destined for public instruction, that reserve, which I have never met with in other countries; for I have, except in France, uniformly obtained all the information which I applied for.

It will, nevertheless, be easy for me to lay before my reader, the particulars which he here looks for, by making an extract from " Instructions on the most proper Means of securing the Increase of Merino Sheep, as well as the Preservation of this Race in all its Purity," drawn up by Gilbert. This skilful agriculturist was specially charged with the care of the Rambouillet flock; and it is in strict conformity to his directions, that this flock has been managed to the present moment.

Before I enter, however, upon his instructions, I shall make some observations, which are the result of a temporary residence at Rambouillet, for the purpose of learning all I could upon the subject.

The Rambouillet flock receives abundant food during all the seasons of the year; in consequence of which the animals are much increased as to size, a fact fully proved by a comparison with those bought by Gilbert in Spain. It is not uncommon to see at Rambouillet rams of twenty-five to twenty-six inches high. The average weight of wool from each sheep is 2 lbs. more than from those bred in Spain, which have been a year upon the same farm.

The sheep are taken to the pastures of natural grass or clover; and sometimes mown lucern is given to them. During winter they are fed on hay, aftermath, lucern, trefoil, vetches, and chaff. They are allowed as much aliment as they will eat. The straw, put into the racks, only serves to amuse them; and being gradually drawn to the ground, becomes their litter.

Some animals are allowed half a pound of oats per day each. The quantity of grain allotted to the flock is considerable; and the individuals which go from Rambouillet to other farms, are in danger of thereby suffering; for it is dangerous to pass from this high treatment to the system pursued even by our most liberal flock-masters. Those, therefore, who make purchases at Rambouillet, should gradually lead them from this species of food to the one which they must ultimately adopt.

It is much to be wished that the use of corn were omitted at Rambouillet in favour of roots, such as potatoes, beets, &c. I have mentioned, in the course of this work, various reasons for preferring root-crops to grain.

The sheep are caused to walk about daily for an hour or two during winter. They are folded in summer from the middle of July to the end of October.

Three flocks are formed; one consisting of the rams, another of the ewes, and a third of the lambs. The latter remain with their mothers during the day; but are separated for the night, and weaned at the age of five or six months.

The sheep-houses at Rambouillet admit a

free circulation of air. The racks, which are better constructed than those generally to be found in the country, would boast still greater advantages, if made on the Northern plan, recorded in this work.

Pasturage at Rambouillet.

Among several fields of sound soil, possessing a diversity of hill and dale, and consequently adapted to sheep, the park at Rambouillet also contains other closes, flat and confined; several are cold, and some wet. The use of these lands is regulated according to the season, the temperature, the hour, the food given to the animals in their houses, and various other circumstances; by which the dangers, naturally to be expected from a less cautious system, are avoided. There is, for instance, such a pasture, which the flock never enters directly after leaving the cots, and such a one, where the sheep must not remain, but gently pass through. To one they are not conducted except in damp weather; to another, except in great droughts. This pasture may be used in a morning; that must not be stocked till afternoon. The more farmers will reflect, and instruct their shepherds as to the effect of humidity on sheep, the more successful will they be, though they may not occupy land of the most favourable kind.

Food.

The Merino breed accommodates itself to all plants, which are relished by other sheep. I fancy, too, that I have observed (and the shepherds at Rambouillet confirm the justice of the remark) an inclination on the part of the Spanish sheep to eat herbs, which the native breeds of France refuse. It does not enter into the plan of this instruction to point out all the substances, which may be used as sheep-fodder; suffice it to say that lucern, clover, sainfoin, upland hay, and, above all, the aftermath of well-dried lucern and clover, particularly suit the Merino breed of sheep.

During the season, a few oats should be given to the rams. They impart vigour, and have great effect on the production; which in size and constitution, as well as in wool, bear resemblance to the father and mother, in proportion to the superiority of vigour possessed by either. It is particularly in the alliance of Merino rams with native ewes, that this attention is of great importance.

A month before lambing-time, it is a good

plan to give the ewes a little bran, or a few oats, peas, beans, or other kind of grain; and this support should be allowed for a month after the birth, or more; unless the ewes can find an abundant supply of food in the fields, or are allowed it at home in another shape. A few oats should, likewise, be offered to the lambs, when old enough to eat them; and breeders should not be afraid of the expense attendant upon this; as they are amply rewarded in the beauty and value of the animals thus reared. In other respects, these supplies of oats, bran, &c. bear reference to the quality of the pasturage. If the latter be abundant and substantial, little aid is necessary; if otherwise, it is indispensable.

The use of salt, very little known in France, produces on sheep in general, but especially those which carry fine wool, excellent effects; and farmers cannot be too strongly urged to adopt it. Half an ounce is, at Rambouillet, given daily to each animal in a little bran or corn. It may also be given alone; and sheep are extremely fond of it.

Water.

In many districts, sheep are never allowed water, and it is difficult to imagine a system more disastrous. Fine-woolled animals should have water every day; and if conducted to it in such a manner as neither to have a dread of the shepherd nor his dogs, there is no reason to fear that they will drink to excess.

Clear, gentle, running streams claim a preference; but in every country, such must be used as are at hand. It is only necessary to be observed, that if no place can be resorted to but such as is corrupted by the moisture from the dunghill, it would be better to give the water in the troughs, or in buckets; which should be stationed in the houses all the time that the animals are confined at home by the inclemency of the atmosphere.

Shelter.

Continued rains being infinitely more prejudicial to sheep than frost, it has been thought sufficient to protect them from the effects of the former, and sheds have accordingly been built. These may, certainly, answer the purpose well enough; but I do not hesitate in giving a preference to houses, spacious enough to prevent any crowding, high enough to be wholesome, and sufficiently provided with apertures for a free circulation of air. If such sheep-houses be built on a dry soil; if there be attached to them a yard or court sufficiently large for the animals to take exercise when instinct directs it; and if proper attention be paid as to the renewal of litter; there can be no doubt that such shelter is the most secure, the most commodious, the most healthy which can be obtained, in every place, and for every season.

Folding.

Opinions are no more fixed with respect to folding than to cotting; and the reason is that people always wish to lay down an universal plan, when it is obvious that variations must ensue from local circumstances. Folding may be conveniently and advantageously resorted to on all lands which are perfectly sound, provided the practice is not adopted till after the period of cold and heavy rains; remembering too not to expose the sheep directly after shearing, and allowing them shelter whenever a storm, or heavy rain takes place.

For want of these precautions arise the rheums, to which sheep are so liable, the obstinate discharge from the nostrils known by the name of glanders, and many other acci-

dents, which are the effect of arrested perspiration, consequent upon folding.

Pasturage.

When a flock has passed the night in a sheep-
cot, or the fold, it is of the greatest importance
not to let them go abroad before the dew has entirely disappeared. Few shepherds are attentive
enough in this respect. Fearing that their
flocks may suffer from the effects of hunger,
they drive them abroad at an early hour, and
lose them. I have often observed that sheep,
left to their own free will in the fields, never
eat herbage, which is wet; but this is not
the case with animals just released after a
night's confinement. Pressed by hunger, they
devour with avidity the plants, which are covered with dew; and this nourishment, by re-
laxing the fibres, accelerates the fattening of
the animal; but it is a deceptious condition
and soon followed by the rot. The conduct
prescribed is, therefore, most particularly indispensable to breeding stock. Be it remembered too, that herbage moistened from any
cause, may produce, more or less, the same
effect as if it were covered with dew.

If compelled to take a flock abroad during

damp weather, it should always be conducted to the most elevated land, among broom and heath, on spots the least exposed; and it should have had the appetite in a great degree appeased by a sufficiency of food before its release.

Low and marshy lands, such as are covered with water during winter, yet dry in summer, ought to be never used for sheep-pasture. If compelled to the measure, the animals should certainly not be admitted till towards the middle of the day, when the ground is perfectly dry. The further precaution should be used of never suffering the flock to remain there but for a short time.

In great heats, it is necessary to take the sheep from pasturage during the sultry part of the day; and to afford them shelter, either under trees, or in their houses, the windows of which, in the latter case, should be left open only on the side, which is not immediately exposed to the rays of the sun.

It may be considered as an established fact, and acted upon as a general rule, that a moderate temperature is the one best adapted to the general nature of sheep, whether reference be made to their health, or to the excellence of their fleece. An intelligent shepherd, thoroughly imbued with this conviction, will soon discover the best mode of securing the preservation of his charge.

The pastures, which are richest, and most abundant in herbage, are those, which should always be looked to with an eye of suspicion. It is extremely dangerous to let sheep feed on artificial grasses without great attention. Lucern, and clover more particularly, cause sheep, and indeed other animals to be *hoven* as it is termed, and thereby to perish in a few hours; and still more especially if such herbage be wet. Too much care cannot, therefore, be taken to avoid this sort of pasturage; and if compelled to use it, the sheep should be gently driven through it, bringing them again and again during the day; but never suffering them to remain in it beyond a few minutes at a time.

If, in spite of this precaution, some animals are hoven, no time must be lost in causing them to run, or throwing them into water, or giving them half a glass of oil; and if, after these expedients, the symptoms do not abate, a penknife must be plunged into the paunch on the left side, immediately under the short ribs. By this incision, an evacuation of the air contained in the stomach is produced, or the food, of which the animal has taken too great a quantity, is disengaged. In order to facilitate this evacuation, as it becomes desirable, a pipe made of reed, elder, or other material, about, as thick as a finger, should be introduced by means of the aperture. The wound may then be left to Nature.

Coupling.

An opinion prevails, which does not appear to me less erroneous because it is general, that in every species of animals, those, which are young, are always the most productive, and supply the most desirable offspring. This would be incontestable if, by young males, were understood adult animals, or such as had reach-ed the proper period of generation; but these are precisely the last, which are regarded as proper for reproduction. Against this opinion, the first elements of sound philosophy may be adduced; and its only foundation is the very abuse of the young animals, which are objec-tionable. It is evident that a ram, eight or nine months old, and yet employed for an hundred or an hundred and fifty ewes, of which I know several instances, will not be fit for generation in the following year; or, at least, that he will not be so fit for it as a younger ram; but if these animals be not used till they have nearly finished their growth, that is, till they touch upon the end of their second year, there can be no doubt but that they will be as vigorous as shearlings, that they will beget a better offspring, and continue so to do till they have reached the age of six or seven years, provided no more than thirty or forty ewes are annually allotted to them. The productions of too young a ram ever tend towards degeneration.

It is, perhaps, of still greater importance to wait for the adult age of the ewes. They are capable of being employed at ten or eleven months old, and they may produce good stock enough at eighteen or twenty months old; but if it be wished to obtain choice animals, and such as are eminent both with regard to fleece and carcase, it is necessary to prevent communication between the rams and ewes till they are two years and a half old; at least unless they are previously very vigorous, and have reached their full growth. If any of the younger reserve should, by chance, have taken the ram, it is most advisable to deprive them of their lambs immediately after birth; and these lambs should be reared with cow's or goat's milk, unless there are other ewes, to which they can be assigned: a method always more desirable. Experience having proved that suckling is always more fatiguing than gestation, the young ewes, which have received the ram at too early an age, will, by this removal of their lambs, be no worse for the connexion which has taken place. Hence, if a breeder wish to arrive at perfection by rapid strides, he may employ shearling ewes with safety, provided he takes care to have good nurses of some common breed, whose productions are consigned to the butcher, or provided he takes care that the high-bred lambs are reared by milk from the cow or the goat. *

When rams and ewes have been thus treated, they may, without inconvenience, be employed, the former till seven or eight years old, the latter till eleven or twelve. There still exist at Rambouillet ewes, which arrived from Spain

---

* Buffon recommends too that breeders will wait thus long before the animals are used; but general experience has surely proved, throughout Britain, that it is quite unnecessary; and the delay would be a great drawback on the profits of a farm.　　　　T.

sixteen years ago; they were then two or three years old, and even now produce good lambs. It is necessary, however, to observe that this longevity is peculiar to the Merino race; at least, breeds of France decline and die much sooner.

There is as great a diversity of opinion respecting the most advantageous period for allowing the ram's access, as there is with regard to the age of the animals employed. On most farms the rams, or rather the ram (for it is seldom that more than one is employed) remains throughout the year with the ewes. In consequence of this, all are generally impregnated; but the lambs fall at different periods, which is very inconvenient for the shepherd; and the rams enervate themselves so much that they are obliged to be annually changed. It is a general consequence too, that the lambs drop during the coldest season of the year, and when the pastures supply least food; a formidable inconvenience, unless care is taken to have an abundant supply of fodder, or the still better support of turnips, beets, cabbages, &c. which all persons should abundantly provide themselves with, if desirous to rear sheep in perfection.

The intention of Nature is that the ewes

should be served when they first feel desire for it; and if particular considerations induce any one to postpone the period, it would be erroneous to suppose that this can be done to any great extent, without much inconvenience.

Though the ewes may have a renewal of desire a fortnight, a month, or two months after the first heat has subsided, it is by no means certain that fecundity will be equally obtained; and that they will, on the second or third return of impulse, be in that condition which insures the strength and good constitution of the fœtus. I at least have remarked, and a hundred others have made the same observation, that if the ewes be put to the ram long after their first inclination, many prove barren; and it is an opinion, founded on general experience, that the lambs first born are always the most vigorous, and always attain a greater size than those afterwards produced.

Analogy comes likewise in aid of this opinion. The brood-mares, which have not their heat satisfied in the spring, often feel it again in summer and autumn; but if covered at these seasons, they frequently are not pregnant; or if so, their productions are constantly more feeble, and in other respects inferior to the foals, which were got at the earlier period of the year.

These observations may suffice to prove that the time for copulation is dependent on local circumstances, and that no general rule can be laid down in this respect; but that the views of Nature should not be opposed unless for purposes palpably desirable, since this can never be done without attendant disadvantages; and that this can always be better done when, instead of relying on the uncertain resource of winter pasturage, an abundant supply of green crops is grown.

It is by the latter valuable culture too that the general unfortunate consequences of a change from green to dry food, and *vice versâ*, are avoided.

Weaning.

Abundant and wholesome food being the essential cause of improvement in sheep, and all the other attentions necessary for the ewes and lambs of the Merino race being simply the same as for those of the native breeds, I pass to the period of weaning; which, as it regards the restoration of vigour in the dams, and consequent renewal of stock, will demand particular precautions.

The lambs should not suck their mothers more than five, or at the utmost six months. At this period it is necessary not only to separate them from the ewes, but to make a division of the ram and ewe-lambs; otherwise, the former would be enervated by trying to serve the latter, and some of these would be impregnated. Both would continue small and ill-formed, their production would be still more so, and degeneration would soon be complete. Too much care, therefore, cannot be taken to prevent the connexion of rams and ewes till they have attained their full growth, which can never be better effected than by forming two flocks, the one entirely consisting of males, the other of females. It is also of great importance to form a third during the season of generation, composed entirely of the females not intended to be used.

Occupiers of land, who have farms near each other, will act wisely in having only one breed upon each, and the same advantages may be procured by maintaining a good understanding with neighbours. It is still better when any one possesses inclosures, in which the different classes can be kept without danger of inter-mixture. These smaller fields are so advan-tageous and save so much expense ultimately, that we cannot too strongly urge farmers to make them, which is almost at every place easy.

It is important towards the success of wean-ing that it should be gradual, for if effected too suddenly, the milk causes disease in the udders of the dams, and the lambs, by so immediate a transition to dry food, are sensibly affected.

Castration.

There are many modes of castration, all of which are good, and about equal as to merit. I shall, therefore, only treat of it, with regard to rams of the 'mixed breed, pointing out how far they may be considered as approaching pu-rity, and fit for generation.

It sometimes happens, but very seldom, that, even in the first cross, a sheep will so much resemble the sire as to make a very sensible dis-tinction between them difficult. At first, therefore, it would seem that a first-cross ram, possessing these perfections, might be employed for purposes of amelioration; but it is a truth, established by a thousand facts, in every class of animals, that the offspring will sometimes bear a greater resemblance to the grandfather and even great-grandfather than to the sire himself; consequently, as the maternal ancestors of this mixed product were of a common breed, it is much to be feared that the perfections of the individual might not be conveyed to his descendants.

This inconvenience, of the greatest importance in an enterprise of improvement, should be continually held in dread till the animals have reached four crosses. If they then possess all the qualities of the pure race, combined with the form, which we look for in a stallion, they may, without hesitation, be used; but till then, however promising they may appear, it is right to take away the organs of generation, or, at all events, remove them from the flock, before they are in a state of reproduction; for if allowed to be together, it is highly probable that some of the ewes will be served by this objectionable stock, and the object in view thereby retarded.

That these animals of the mixed breed, including even those of only one cross, are infinitely more proper to be used than sheep of a common kind, and that they are also capable of imparting a most sensible improvement to coarse-woolled breeds, is an incontestable truth; but the only consequence, to be deduced from this observation, is that the mixed breed should be used as long as it is impossible to procure the pure race, and no longer. It is, nevertheless, not to be denied that a powerful consideration offers itself in favour of the mixed sheep; namely, the wish, felt by all enlightened agriculturists, that the repugnance, exhibited towards the adoption of animals not possessing the form, to which the idea of beauty is generally attached, should be, as soon as possible, banished. If all the rams of the mixed race, bred in France during the last twenty years, had been preserved for propagation, it is certain that there would now be hardly a flock, which did not possess some portion of Merino blood. The eye would have been accustomed, by imperceptible degrees, to the form of these sheep; the opinion, with regard to the true character of beauty, would have been changed in favour of the pure Spanish ram; and this model would have been, at length, fixed upon as the object, towards which the general view should be directed; for the mind naturally fixes on objects, which supply sure and easy results, in preference to those, which offer great advantages, but dependent upon eventual circumstances.

The general interest then, as well as the interest of individuals, points out a preservation of the mixed breed; but care must be taken that no ram of this kind has communication with pure ewes; since such a system would be retrograde.

Shearing.

The finer and closer that the fleece is, and the more regularly that it is spread over the surface of the body, the more necessary is it to preserve from the inclemency of the atmosphere the animals, from which this valuable covering has been lately taken. Excessive heat is not less to be dreaded, under such circumstances, than cold and moisture. The most moderate temperature is that, which should be adopted for sheep during a few of the first days, after shearing. If they were before abroad, it now becomes necessary to place them under sheds, or in sheep-houses perfectly admitting the air; but if the latter be low and confined, it is much better to let the animals remain out of doors.

On the same principle, the method of washing the wool of the Spanish breed upon the backs of the sheep, should be opposed, and in fact it ought to be proscribed with respect to every race; for it is attended with no advantage whatever, and is, on the contrary, fraught with much inconvenience. No great knowledge is necessary to perceive what must be the effects of leaving a fleece soaked with water, to dry on the body of an animal, to which humidity is more fatal than any thing. The quality and preservation of the wool do not less call for the abolition of this custom than the health of the sheep ;—a custom, in many places followed merely because farmers continue in the track, which they were originally taught to pursue. *

* The above remarks equally apply to all sheep in Britain; but more especially the Merino race, and its crosses. When Mr. Tollet published an account of his flock in No. 243 of Young's Annals, he said: " From the quantity of yolk in the Merino fleece, it would be difficult to wash the wool before shearing, at least to produce any good effect. The closeness of the fleece preventing the dirt from getting into it, no inconvenience arises in the shearing: and wash-

Amputation of the Horns and Tail.

The horns, which Nature allotted to the ram for his defence, not only become useless,

ing it would only tease the animal, to answer no good end; as the manufacturer would have nearly as much trouble in scouring the fleeces afterwards." Dr. Parry also adduces many good reasons on the same side; but the wool-buyers object to this change, stating that they can form no just estimate of the real weight, unless the general custom of the country be pursued. The Doctor, however, is convinced that the waste, occasioned by washing on the back of the animal, is very variable, and this he substantiates by abundant proof; consequently the buyers know, in reality, no more after this incommoding and sometimes dangerous process, than they would know, if it were omitted. In town, it appears that they are not so averse to a purchase of wool in the yolk, as they are generally found to be in the country; for Lord Somerville states in a letter, which I lately had the honour to receive from him, that three packs of wool, from his Lordship's mixed breed, (each containing Two Hundred and Forty Pounds) have been sold by auction in London at Five Shillings and Two Pence per pound *in the grease*, which is probably equal to Eleven Shillings per pound, when scoured to a state of perfect cleanness for manufacture. This is the highest price for wool of the *mixed* breed, which has yet come to my knowledge; and its real value could certainly be ascertained in no fairer way than by public sale. T.

but even inconvenient and hurtful to him, in a state of domestication. They forbid the approach of his head between the interstices of the rack for the purpose of picking the straw, in which he is nice, and for culling the few ears of corn or choicer herbage mingled with it ; they sometimes wound the ewes as they pass through gates ; and they are not unfrequently fatal to the rams themselves during the combats, which take place among them. There are two modes of amputating the horns; one being by the saw, and the other by a chisel. In the first case, a handsaw of very nice edge is employed, and those made in England are best adapted to the operation. A man holds the head of the ram firmly, while another makes the amputation, which is very soon done, if the operator be skilful.

The amputation by the chisel, which is practised by the Spaniards, is not so simple. A ditch is made, which, in length and breadth, is about the size of a sheep, while in depth it is about five or six inches ; and another is formed rather smaller than the former, at one end of it, the two, thus forming together, a cross. In the last ditch, which is still more shallow than the other, a large plank is fixed to support the

Q

head of the ram when he lies upon his back in the one, which forms the trunk of the cross. A man stretches his body over that of the ram, and holds the head forcibly upon the board with one hand, while with the other he holds a long and large chisel weighing four or five pounds; which is then fixed on the horn, and another man striking it with a mallet, the part intended to be removed, flies off at one or two blows. The preparation, necessary for this method, makes the saw worthy of decided preference.

The operation is generally performed when the animal is one year old. It is not unusual, when the horns push forward again, that some parts of the head are touched by them, in which case they become so troublesome, and sometimes injurious, that a second amputation is necessary. *

* Mr. Tollet has adopted the still safer method of breed_ing rams without horns, which experiment he was enabled to try by a polled ram_lamb being produced among the rest of his pure stock. This animal proved to be of good form, and to have an excellent fleece. His male offspring were partly polled, and partly otherwise; but by constantly breeding from the former, and rejecting the rest, Mr. T. has succeeded in almost banishing the horn from his flock. I have sometimes thought that these polled sheep, though yielding wool of admirable quality, do not, in general, carry

The tail of a sheep is an appendage, which is almost useless, and generally inconvenient; for

quite so large a quantity of it as the horned stock, and I have heard the observation made by others. Of three pure Merino rams, which I bought of Mr. Tollet at different times, one is hornless; and though superior in fineness of wool, compactness of frame, and lightness of bone, his fleece is not proportionate in weight to those of the other two. There are, however, instances, opposed to this; and I particularly remember one in Mr. Tollet's flock, from which a polled ram was, at his shearing last year, let to Lord Bradford at fifty guineas for the season; and on that day he yielded 10 lbs. of excellent wool, washed in the usual manner on the back of the animal. Should it eventually prove that no general diminution takes place in the weight of wool carried by polled rams, it will, I conceive, be desirable, on many accounts, to rid this valuable race of the incumbrance. Those sheep, which are thus armed with power of annoyance, are apparently more quarrelsome than the other. The fly too is particularly troublesome by its usual attack on the root of the horn; and a remark has been made that the animals, bred without horns, are almost always more level on the top. It would, however, be far from a desirable measure to discard the horn from all breeds of sheep, even if possible; for more than one manufacturer has mentioned to me the important fact that the combs, used in preparing wool for worsted fabrics, cannot be properly made without sheep's horn; every other hitherto thus applied, having proved very inferior.

T.

it becomes occasionally loaded with dung, a
great proportion of which it transfers to the
fleece. The breeders in England, Spain, and
in almost every country, which attaches importance to the amelioration of wool, are very
careful to dock their sheep; they even go so
far, in some instances, as to think that the
practice contributes to the barrelled appearance
of the animal. Waving, however, this opinion,
which is probably an optical illusion, it is certain that the dismissal of the tail is attended
with advantages. The operation should be performed when the animals are three or four
months old; the tail should be cut off to the
length of three or four inches; and there is
some danger in docking closer.

CHAPTER VII.

HOLLAND.

THE mode of treating sheep, pursued in Holland, is dependent on the nature of the country,

and consequently differs from those hitherto
mentioned. Each farmer has some sheep on
his land; but in general the number is small.
These animals may almost be described as left
to themselves; and feed with cattle, horses, or
pigs, in fields, sometimes surrounded by dikes,
at others by hedges. They are never put into
houses, but when the ground is covered either
with rain or snow. It is customary to leave
them too, throughout the year, on the pastures
of North Holland and the Texel, departments remarkable for the number of sheep
kept in them. In this part of the country, the
peasants usually possess from one to three hundred sheep each; while, in other districts, they
content themselves with rearing a dozen, and
rarely more than forty.

It is found that these sheep, principally of
large size, fatten less kindly when a great number are kept in the same pasture, whatever
may be the abundance of herbage; and one
reason for preferring a large kind is, that a certain duty per head, being paid upon sheep of
every description, those of the greatest size are
esteemed least expensive, consequently most
profitable.

I will here describe the remarkable breeds of sheep in Holland, and the Netherlands.

The Friesland race is remarkable for size, beauty of form, and abundant produce, which last description embraces wool, milk, and lambs. I have measured animals, which proved two feet nine inches high, whether from the ground to the top of the withers, or of the rump. Their length from the tip of the nose to the prominent rump-bone was four feet five and a half inches. They are hornless, and the tail, which is nine inches long, is not furnished with wool; in thickness it does not exceed that of the middle finger. The ears are long, even reaching almost eight inches; and the head eleven. They carry a silky kind of wool, the staple of which is fifteen inches, and very fine for its length. They are of a prodigious size, but constantly very lean, and especially at the time they are milked. The fleece weighs sixteen to seventeen pounds. The ewes which have udders as large as that of the goat, are milked twice a day, and give a pint at each meal. They annually produce three, four, and even five lambs.

There is a great resemblance between the

sheep of Friesland and the Texel. They are alike in form, but the latter are rather less, carrying also wool neither quite so fine, nor so long. They are shorter in the leg, and have a short tail, but large, and covered with wool. The ewes afford less milk than those of Friesland, and are also milked twice in the day; this produce, however, makes cheese of excellent quality. These animals have three or four lambs annually.

The sheep of Friesland and the Texel constantly graze on pastures which are marshy, and even sometimes covered with water; without being more subject to disease than the animals, which feed on uplands. The size of these sheep is, doubtless, ascribable to the excellence of the pasture, in which they remain during a great part of the year, and to the substantial food, which is allotted to them in the stables. The two breeds, which I have mentioned, are highly advantageous to the owner in a country like Holland where the herbage is remarkably abundant and nutritive; but on less fertile pastures they would degenerate, and make a smaller return than those of moderate size, which can find sufficient food on tolerable land.

I have met with this race in Holland, Holstein, the coast of the Baltic, as also in Sweden and Denmark; but the animals had degenerated more or less as their pasture was more or less abundant.

It appears that there is, in Ireland, a breed of sheep, which is even larger than that of Friesland, and yields wool longer as well as finer.

Some departments of Holland possess a race with long legs and tails; but inferior in size to the above. Their wool is coarse, and thinly scattered over the body. They are kept in flocks, and deemed preferable to the other breeds for downs and moors.

These sheep yield six to seven pounds of unwashed wool each, and the Texel race eight to nine pounds. The last sells for ten to thirteen pence Dutch, while the other only reaches nine to ten.

The mode of keeping the indigenous breeds of sheep as well as the Merinos, varies in Holland according to local circumstances, or according to the degree of intelligence possessed by the farmers. I shall give an account of Mr. Twent's method.

His Merino sheep are kept, during summer,

on the Downs, which are favourable to health from the dryness of the soil, but which supply only a scanty portion of herbage. It has, nevertheless, been observed that the Merinos, when also put upon low and marshy grounds, do not become diseased, or carry an altered and degraded fleece. The low grounds, which consist of land under tillage, and coppices of oak or elm, are intersected by ditches destined to receive the water, which drains from the soil.

The Merinos are also suffered to graze occasionally on the high-roads; and still more frequently in the woods, where they find abundance of herbage. When cold, they are taken to the houses during the day, and they never stay abroad all night.

They go into the fields every day, unless the snow is deep, in which case they receive at home the same food as is given to the native breeds, viz. hay, beets, turnips, beans, and oats.

Mr. Twent has ascertained that the food given to the indigenous breeds, as well as the pasturage of the country, will perfectly suit the Merinos, and that the latter fatten as kindly as any. He has also remarked, that the Merinos, which live on marshy land, are less subject to the rot than the sheep of Holland.

Another observation, not less interesting, has been made by Mr. Twent for a dozen years. This is, that the leaves of the alder, to which sheep evince a strong partiality in damp weather, preserve them from the rot. Nature appears to have pointed out this preventive; for it is principally in rainy weather that the disease takes place.

Mr. Twent gives no salt to his sheep, which, nevertheless, constantly enjoy good health. During some months he folds them, contrary to the custom of the country.

CHAPTER VIII.

ITALY.

PIEDMONT, in imitation of France, attempted to naturalize the Merino race of sheep and the success, hitherto attained, augurs favourably for the future. The measures, indeed, which have been taken to propagate this valuable breed, afford reason to believe that this part of Italy will soon produce the quantity of fine wool, necessary for its manufactures of cloth.

The details, which I shall here present as to the breeding of Merino sheep in Piedmont, are extracted from a memoir presented to the Academy of Agriculture at Turin.

Since the arrival of the Spanish flock in Piedmont, (1793,) it was resolved to procure the best ewes from the Neapolitan States, Germany, and the vicinity of Padua. In the spring of 1795, these were joined to the Merino flock, which had already produced lambs; and from this time the two breeds had the same care bestowed on them, the same management, and the same food; in consequence of which they alike prospered. The lambs of the pure breed had their tails cut a few days after birth, that no doubt might exist as to their legitimacy; every male of the mixed breed was castrated before six months old; and those of the pure race were not employed till in their third year, when a preference was given to the most beautiful animals.

The flock was divided as follows:

1. Pure rams.
2. Wethers, to be kept till three-shear.
3. Breeding-ewes of all kinds.
4. Younger ewes.
5. Rams of the first and second year.

The flocks are kept, on the beautiful farm of La Mandria, by intelligent shepherds, under the direction of a chief taken from their own class, who has further under his orders, a person in each division, and boys in proportion to the number of each flock.

In the midst of the farming buildings is a quadrangular court. The vestibule, by which access is given to this, contains, on the right and left, lodges for the shepherds. Three large covered sheds, each capable of containing a thousand sheep, form the other three sides of the square. Each of these sides has two large doors opening into the court, and the apertures are also furnished with grates four feet high. The sheds have an interior communication with each other. Large windows in the sides of these wooden buildings, cause a free circulation of air, and facilitate the exhalation of the dung. Mangers are fixed against the wall and in the middle of the buildings almost to their full length, by which all the sheep can eat at once without pressing against each other. Lastly, moveable divisions are at hand either to enlarge or contract, at will, the space allotted to any lot of animals.

At the entrance of the buildings there is a running stream, which supplies a very large reservoir. Eight-hundred acres of land, in the centre of which these buildings are placed, furnish the pasturage and fodder necessary for the flocks, while they continue in the plains, which is from the end of October till the middle of June. At this latter period they are driven to the Alps, where a second spring offers abundant pasturage, and superior to that, which they have quitted. The lambs are not shorn; but retain their fleeces till about eighteen months old.

The other sheep, which have been deprived of their wool before departure, seem to feel still greater delight in these elevated regions; and the rams, before slothful, perform, under this happy influence, the great work of reproduction.

No system of uniting animals can be more advantageous; for vigour, force, and health can only be the result of pure air, excellent food, and free use of the faculties. Hence, abundant as is the nutriment allotted in the plains, these sheep never are in higher order than when they descend from the mountains. The season of generation occupies only a fortnight; at the expiration of which time the rams are formed into a separate flock; and in the month of December a tolerable calculation of the lambs to be expected may be formed by counting the ewes, as abortions are very rare. This system would leave nothing to be wished, if means could be found to prevent those two destructive maladies, the rot and giddiness. Hitherto, the other disorders, incidental to sheep, have made no ravages. Sound and dry pasturage must be resorted to as a guarantee against the former, and a judicious selection of sires against the latter.

The rams are never employed till in their third year, and the ewes in their second. Longer delay would be improper, the animals being then perfectly formed.

Folding is but little practised; experience having proved that housing is much better when there is an abundance of straw to increase the dunghill; but on mountains, where there are no buildings, and straw is scarce, a fold is used throughout the time that they pass over them. When the ground is covered with snow, and when the pastures, from any other cause, do not supply a sufficiency of aliment, the sheep are fed in the houses on the best aftermath. Such is the fertility of Piedmont, and the supply of fodder, that no other crop has yet been resorted to for the use of the flocks. It is intended, however, to try potatoes, of which some are already grown in the plains.

Each lamb is suckled by its mother only, contrary to the custom of Spain; and it is separated from the dam at forty days old; for it is necessary that the weaning should take place some time before the journey to the mountains, in order that the animals may be strong enough to encounter a drift of seven or eight days.

CHAPTER IX.

EXPLANATION OF THE PLATES.

PLATE the First is intended to shew the parts of the sheep, on which the different sorts of Merino wool are grown.

In Spain, the wool is divided into four classes. That of the best quality is called *Rafino*, or Superfine; the second is termed *Fino*, or Fine; the next *Tercero*, or Third kind; and the last *Cahida*, or Refuse.

1. The *Rafino*, or very best wool, is taken from the withers, the back, the rump, the lateral parts of the neck, the sides, and the shoulders.

2. The *Fino* is from the thighs, the belly, and upper part of the neck.

3. The *Tercero* is that growing on the cheeks, throat, breast, the upper part of the fore-legs, and the lower part of the thighs.

4. In the last class, or *Cahida*, is included the wool growing on the upper part of the head, on the legs, on the tail, on that part of the breech, which is soiled by the animal's dung; also the wool growing under the belly, and

between the thighs, the small portions, which fall during the operation of clipping, and other stray locks.

Plate the Second is meant to elucidate the mode of numbering a flock.

As the formation of a flock should not be left to chance, but be directed by experience founded upon proper calculations, it is necessary that the breeder should know every sheep belonging to him, and be able to trace every improving generation to its original source.

This knowledge is useful, whether applied to the pure race only, to the crosses of the Merino ram on native breeds, or to the mixture of any different kinds among each other.

In the first case, it becomes necessary not only to maintain the breed, but gradually to ameliorate it more and more, as well with respect to the fineness and length of wool, as in other respects, such as carcase, aptitude to fatten, &c. It becomes requisite, therefore, to know the animals gifted with the qualities desired, or producing the best stock, &c. It will be easy through these means, to form connexions in the flock, by which sheep will be obtained, superior to their progenitors.

The mode of numbering, here submitted to

notice, will be equally useful when the object is to combine two or more breeds with each other; for it will point out which are, in each race, the animals most advantageous in crossing; it will shew which is the rising stock of greatest promise, and which ought to be rejected. This, indeed, may be effected by any method of marking, if durable; but in obtaining such a distinction lies the difficulty. It should be a system, which makes the breeder clear of error, and at the same time easy in its execution. I shall merely state my method, and any one can change, or modify it according to his own liking.

It is customary to mark horses by the application of a hot iron upon the thigh; but the fleece, in which a sheep is clothed, does not allow of this, unless, indeed, the iron be used on the forehead, cheeks, or horns of the animal; and in this case, it should only be for the purpose of distinguishing any particular class. But it is further necessary to give each animal its own peculiar mark. Some persons make use of small pieces of metal, on which the numbers are engraved. These have holes made in them, and are attached to the ear or tail by packthread, or wire. This mode is rather expen-

5

sive, and cannot be adopted in all cases. It is attended too with the inconvenience of the marks being sometimes lost, which is very disagreeable to those, who are desirous of keeping an exact account, and of ascertaining the results of certain experiments. The easiest method, and that, by which the end in view is best attained, consists in marking the ears of the animals by incision.

If a flock of pure Merino sheep, and another of the mixed breed be kept upon the same farm, it will be necessary, for the purpose of avoiding mistakes, to give all the individuals of the first flock a mark, by which they may be securely distinguished from the second.

In this case, a small iron, previously made hot, must be applied to the face of these animals, and leave the impression of a letter or a number.

Next, every individual should be ear-marked, and two Roman figures, viz. I and V, will suffice to form a series of numbers as far as 199. On reaching this, a new series will take place, by marking the animals with a different brand, or upon a different part, and thus a flock may be accurately distinguished to any extent.

The marks may be made either on the upper

or under edge of the ear, the left pointing out the tens, and the right the units. Some prefer the under part of the ear, because the upper edge better protects this organ from rain, and other atmospheric injury.

The number I will designate the animals as far as four inclusive, and to mark five a portion of the ear's edge is cut out in the shape of V. The numbers immediately following, as far as nine inclusive, will be indicated by the following method:

No. 6,	- -	VI.
No. 7,	- -	VII.
No. 8,	- -	IV.
No. 9,	- -	IIV.

The marks, which, on the right ear, express units, will, on the left, express a corresponding number of tens.

Another system of numbering may be adopted, by which it will be easy to know, at the first examination, the genealogy of each individual. If a person has a flock, from which he particularly wishes to breed, he will first mark the sheep as we have already mentioned; and he will afterwards distinguish each lamb by the number of its sire, on one ear, and its dam on the other. In this case the upper edge of the ear may serve to mark the units, and the under one the tens. It will be easy to ascertain the series of generation by examining the mouth of each animal. This mode has its advantages, especially when no register is kept.

But here I must recommend a regular book, in which the numbers of every sheep belonging to the breeder, are inserted. Observations will be made in this, to which occasional reference is necessary and satisfactory ; such, for instance, as relate to the union of particular animals in the pure breed, or the selection for crosses, and interesting results from these, &c. A careful breeder, and one, who wishes to attain perfection in his art, will note the good or bad qualities predominant in certain individuals, their general state of health, their superiority of fleece, or the reverse, &c. By such attentive observation, it will be easy to select those animals for propagation every year, which possess the most desirable qualities, and thus ultimately to obtain the greatest advantages, which a valuable flock can afford.

THE END.

THE FOLLOWING BOOKS on SHEEP, WOOL, &c.

ARE LATELY PUBLISHED BY J. HARDING,

36, ST. JAMES's-STREET, LONDON.

1. FACTS and OBSERVATIONS relative to SHEEP, WOOL, PLOUGHS, and OXEN; in which the Importance of Improving the Short-woolled Breeds of Sheep by a mixture of the Merino Blood, is demonstrated from actual Practice; together with some Remarks on the Advantages which have been derived to the Author's Flock from the Use of Salt, &c. &c.
By Lord Somerville.
Third Edition enlarged, with 10 Plates, price 8s.

2. OBSERVATIONS on the INFLUENCE of SOIL and CLIMATE upon WOOL. From which is deduced a certain and easy Method of improving the Quality of English Clothing Wool, and preserving the Health of Sheep; with Hints for the Management of Sheep after Shearing; an Inquiry into the Structure, Growth, and Formation of Wool and Hair; and Remarks on the Means by which the Spanish Breed of Sheep may be made to preserve the best Qualities of its Fleece unchanged in different Climates.
By Robert Bakewell.
With occasional Notes and Remarks, by the Right Hon. Lord Somerville. Octavo, price 6s. 6d.

Principal Contents of the Work.

Chap. 1.—On the soft and hard Qualities of Wool, and the great Difference in the Value of Cloth made from these Wools, although each sort may be equally fine. On the Distinction between Hair and Wool, &c.—Chap. 2. On the Causes which produce the hard Quality of Wool in many Parts of this Island.—Chap. 3. On the Means by which the soft Quality of Wool may be preserved, in every Situation, and the Effects of Soil and Climate counteracted, where they are unfavourable to this Quality; on the Preservation of Sheep by the same Means; from Cutaneous Distempers; from the inclemency of Climate; from sudden Change of Temperature after Shearing.—Chap. 4. Improved Method of Washing Sheep; on the Means of Preventing the Deterioration of Wool in the Spanish Breed of Sheep; on the Qualities and Defects of the finest Fleeces from Saxony; on the Influence which Pasture, Heat, Cold, and Moisture, have upon the Staple; different Effects of the Climate of Spain and England upon the Fleece. —Chap. 5. On the Formation of Wool, Hair, and Silk; Microscopical Observation on the Structure of each; on the felting Quality of Wool and Hair; Causes of the Variations in the Fineness of the Fibres of Hair and Wool; the Joint, or Break in the Staple; Cotted Fleeces; coarse Hemp Hairs; Effect of Light upon Wool near the Tropics; Asiatic and American Wool-bearing Animals; an Illustration from the Structure and Growth of Feathers; Hints offered by Nature for the Improvement of Wool.—Chap. 6. On the Effects of Lime-stone and Chalk on Wool in different states of Induration; Advantages of the Merino Breed of Sheep; on the Natural Casting of the Fleece; Injurious Effects of suffering Wool to remain piled in an unwashed state; the Necessity of Shelter for Sheep and Cattle in this Climate.—Chap. 7. M. Lasteyrie on the Construction of Sheep-houses; judicious Precautions of the Ancients to preserve the Whiteness of their Wool; an Instance of the Effect of the Climate of St. Domingo on Sheep; on Annual and Biennial Shearing.

---

3. An ESSAY on WOOL: containing a particular Account of the English Fleece, with Hints for its Improvement. Addressed to the Grower, Dealer, and Manufacturer,

By J. LUCCOCK,

Wool-stapler, of Leeds. Price 7s.

Contents of the Work.

Introduction. Wool claims attention from Farmers, Land-owners, Manufacturers, the Public.

Section 1. General Account of Wool. Not always produced by Sheep. Description of their Coats. Wool resembles Hair. Culture improves the Fleece. At first changed its Colour. Is connected with the Arts—and most visible in the Furs of Sheep.

Section 2. Cultivated Wool. What the Author intends by it. Circumstances which promote the Culture of Wool and have a remote Effect. Accident and Caprice. State of Society. Art of Husbandry. Public Taste. Comfort of Woollen Clothing. Discovery of the Felting Quality. Application of Spinning and Weaving. Invention of Machines. Art of Dyeing. Institution of separate Manufactures. The Culture of Wool suspended when Manufactures declined, but renewed when they revived. Evinced in the History of Persia and Rome, Venice and Spain. The Netherlands, England, France, Ireland, Sweden, and Russia.—Promoted by the Invention of New Articles, viz. Worsteds, Tapestry, Hosiery, and Hats. Causes which act immediately upon the Fleece.—General Principles enumerated—Constitution of the Sheep—Temperature—Dryness and Moisture—Pasture Herbage—Soil—Shepherd's Attention to Health and Cleanliness, to Uniformity of the Pile, to its Purity in respect to Yolk, Pitch, Excrement, Moisture, Moats, Time and Manner of Shearing—Fatigue, Cotting—Luxuriant Feeding—Taring the Staple—Age of the Sheep —Scab, Winter-stain and Felting.

Section 3. Essential Qualities of Wool. The Growers complain of Ignorance—The Business of a Stapler described—Different Instruments require different Properties in Wool—The Card and the Comb described, with the Qualities suited to each—Length, Curvature, Pliability, Toughness, Felting property, Softness, Colour, Relative Weight—Smell—Trueness of Hair, &c.

Section IV. Wool of England—All of it imperfect—Divided into two Kinds—Districts producing Long Wool—Description of their Sheep—Sometimes ambiguous—Difficult to ascertain the Quantity of Stock. Long Wool. *Teeswater District*—Durham, Yorkshire. *Lincoln District*—Holderness, Lincolnshire, Norfolk, Cambridgeshire, Huntingdonshire. *Leicester District*—Leicestershire, Northamptonshire, Rutlandshire, Warwickshire, Staffordshire, Customs of the Lincoln and Leicestershire Districts. *Kent District*—Romney Marsh, Marshes of the Thames and the Coast, Customs and Markets. *Devonshire District, Cotteswold District.* Total Quantity of Long Wool. Short Wool. Sheep which yield it. *Norfolk District*—Norfolk, Suffolk, Cambridgeshire, Huntingdonshire, Bedfordshire, Essex, Customs, of this District. *South-Down District*—Sussex, Markets and Customs, Kent, Hampshire, Surry. *Wiltshire District*—Wiltshire, Berkshire, Oxfordshire, Buckinghamshire, Hertfordshire, Middlesex. *Western District*—Dorsetshire, Devonshire, Cornwall, Somersetshire, Agricultural Societies, Gloucestershire. *Hereford District*—Herefordshire, Customs compared with those of Norfolk, Monmouthshire, Worcestershire. *District, with various Breeds of Sheep*—Shropshire Staffordshire, Warwickshire, Leicestershire, Lincolnshire, Nottinghamshire. *Heath District*—Derbyshire, Cheshire, Lancashire, Yorkshire, West-Riding, East-Riding, North-Riding, Westmoreland, Cumberland, Durham, Northumberland, North Wales, South Wales, Isle of Man.

Section V. Concluding Remarks. Total Quantity of Fleece, Skin; and Lamb's Wool—Often estimated erroneously—Number of Sheep diminishing—English Wool susceptible of great Improvement—Average Quality—Modes of raising it—Relative and real Value—Price—Common Mode of estimating the Consumption fallacious—No Combination among Staplers.

---

4. A LETTER to the Most Noble the MARQUIS of TITCHFIELD, President of the Newark Agricultural Society, on the Practicability and Importance of introducing the Merino Breed of Sheep, extensively upon the Forest Farms of Nottinghamshire. By BENJAMIN THOMPSON, Esq. of Redhill Lodge. Octavo, 1s.

---

5. OBSERVATIONS on LIVE STOCK; containing Hints for Choosing and Improving the best Breeds of Domestic Animals. 4to Edition, with an Appendix on the Merino Breed of Sheep. By GEO. CULLEY, Esq. With Plates, 8vo. 7s.

---

6. A GENERAL TREATISE on CATTLE, the OX, the SHEEP, and the SWINE, comprehending their Breeding, Management, Improvement, and Diseases. By JOHN LAWRENCE. Second Edition price 19s.

---

7. FACTS and EXPERIMENTS relating to the Use of SUGAR; pointing out a more Œconomical Mode of Feeding Cattle, than is pursued at present, with Hints for the Cultivation of Waste Lands; shewing the possibility of Lowering the present High Price of Butchers Meat; Encreasing our Supply of Corn; and Improving the Condition of the Lower Orders of Peasantry in Great-Britain and Ireland. Octavo, &c.

Harding and Wright, Printers, St. John's Square, London.